The Staircase

The MIT Press
Cambridge, Massachusetts
London, England

THE STAIRCASE

History and Theories

JOHN TEMPLER

Second printing, 1994
© 1992 Massachusetts Institute of Technology

All rights reserved. No part of this book may be reproduced in any form by any electronic or mechanical means (including photocopying, recording, or information storage and retrieval) without permission in writing from the publisher.

This book was set in Bembo and Copperplate by DEKR Corporation and was printed and bound in the United States of America.

Library of Congress Cataloging-in-Publication Data

Templer, John A.
  The staircase : history and theories / John Templer.
    p. cm.
  Includes bibliographical references and index.
  ISBN 0-262-20082-1
  1. Staircases—History. I. Title.
NA3060.T46 1992
721'.832'09—dc20   91-16960
                                      CIP

Publication of this book has been aided by a grant from the Georgia Tech Foundation and the College of Architecture, Georgia Institute of Technology.

*To Huzur Maharaj Charan Singh Ji
and to my wife, Joan*

Contents

Preface		ix
Acknowledgments		xiii
Introduction		3

1	**Straight Flight Stairs**		13
1.1	Prototype: Climbing Poles and Ladders		13
1.2	Archetype 1: The Architecture of the Internal Stair		19
1.3	The Stair as a Passage		19
1.4	Archetype 2: The Architecture of the External Stair		34
2	**The Helical Stair**		53
2.1	The Defensive Vis		54
2.2	The Helical Stair as an Architectural Device		56
3	**Composite Stairs**		87
3.1	Landing Places		87
3.2	The Dogleg Stair		89
3.3	From Square Newel to Imperial Stair		92
3.4	The Influence of Garden Stairs		102
3.5	Learning from Formal Entry Stairs		113
3.6	The Rise of the Internal Grand Stair		119
3.7	The Baroque Stairs of Honor		128
3.8	The Nineteenth and Twentieth Centuries		160
Notes			171
Glossary			173
References			177
Index			181

Preface

The architect who does not conceive of a staircase as something fantastic is not an artist.
—Gio Ponti

Architects, all idiots; they always forget to put in the stairs.
—Gustave Flaubert

Stairs, ladders, and ramps entered the thesaurus of building components in prehistoric times. Yet even the earliest known and simplest demonstrations are still constructed today. As new variations have developed, these have been absorbed into the vocabulary, usually without displacing other models. The fundamental prosthetic nature of stairs encourages this conceptual longevity, as well as constant rebirth and replication. This functional destiny is so circumscribed by the boundaries imposed by human physical dimensions and gait that the quintessential geometry of the stair has changed more in response to shifting aesthetic, architectonic, and cultural goals than from theoretical or empirical refinement.

The staircase can be a treacherous as well as a beautiful siren, with many facets to its complex overt nature. Historically, the great theoreticians of architecture have recognized the many faces of the staircase, and their treatises evidence a concern for the comfort and safety of those who use stairs. Curiously, the major texts that comprise the history of architectural theory have comparatively little to say about the stair as an aesthetic element. In fact the normative rules for beauty that are typical of so much of the writings of architectural theorists are conspicuously absent from the seminal texts that treat stairs. Many of the great architect-theorists from Vitruvius to Alberti to Blondel and after have thought and written about stairs (and their ideas are discussed in these two volumes). In their writings they treat stairs as architectural elements to be thought of in terms of *firmitas, utilitas,* and *venustas*; order, arrangement, eurythmy, symmetry, propriety, and economy. They think of stairs as the veins and arteries of buildings (Vasari, Scamozzi); as objects of beauty (Palladio); as symbolic (Alberti, Martini, Palladio); as places of danger (Vitruvius, Palladio, Guadet); as places where one should be concerned about the needs of the elderly and infirm (Palladio); about the necessity for good illumination (Vitruvius, Palladio); about the behavior of people on stairs (Leonardo, Palladio); they pontificate about the dimensions of risers and treads, the steepness of staircases, and so on. They also sense that one can think of the stair in some holistic fashion—as both subject and object, as architectural piece and as architecture.

Nineteenth- and twentieth-century specialization has acted to erase or shadow this synoptic view of architecture. For some, today, architecture is essentially engineering (firmness). For others it is function and finance (commodity). For some, it is all and only delight. Some writers suggest that this conceptual minimalization and disintegration of architecture is the inevitable consequence of the scientizing of thought products arising from the philosophies formulated in the Enlightenment and ultimately transposed into materialistic positivism. This position strikes me as romantic, at least, and historically selective and revisionary. It is a curiously materialistic dismissal of the Tao of knowledge. By definition, neither science nor art narrows our learning, even if our views become selective. We alone set our intellectual boundaries. To equate science (but not art, for some reason) with materialism is to equate the spiritual, in the humanistic sense, with the unknown and to give ignorance the status of nobility. This is to glorify a limited and limiting view.

The first chapter of Michel Foucault's *Archaeology of the Human Sciences* (1973) uses Velasquez's painting *Las Meninas* as an archaeological icon to be comprehended only by stripping it of its many layers of meaning. Without understanding the historical moment, the customs of the court, the painter's philosophy (and privileges) and our own, and the roles played by the now mute faces, as he shows us, this painting represents little more than its own dust cover.

Similarly, without a history (in the medical sense), the staircase, chameleonlike, disguises itself to match the interests of the viewer; it is art object, structural idea, manifestation of pomp and manners, behavioral setting, controller of our gait, political icon, legal prescription, poetic fancy, or the locus of an epidemic of cruel and injurious falls. Each disguise is a fragment of its nature that reveals a character but still leaves its essence concealed or fragmentary. To be seduced by only one or two of its masks is to miss the whole play

Preface

of meanings, contexts, and potentialities. It is this partial vision that limits our understanding and inevitable judgments about architecture and building and the elements from which we compose them, leading us, on the one hand, to a mechanistic, insensitive, soulless, and ugly world that claims to be well engineered (in a narrow sense), and, on the other hand, to a world that claims to be poetic, attuned to the ganglia of the human spirit, but remains ignorant of or even disdains the human sciences that can make our art humane, sensitive to our cultures, our differences, our fallibility and imperfections.

Specialization has given us a depth of understanding but threatens our comprehension. Each specialist in our world of design is dismayed that so little of his or her field is held in esteem. Architecture, some say, is form or formal dialectic. It is indecipherable without revealing its social, economic, and political intention, others demand. To the homeless, architecture may be stripped to its functions as building, shelter, and climatic modification. Humanity is still the measure (but not humanity as experience), say others. Without revealing the philosophical theories, architecture is simply a vacuous craft, some insist. And they are all right, in part.

The position I have taken in these two companion volumes is comfortably with Palladio and his followers. It seems fruitless to try to understand architecture, or a piece of architecture, from a single vantage point. That would be to have a partial view (in every sense). The stair offers an extraordinary opportunity to try the opposite approach. To apprehend it, one can probe it from many directions, like the black box of systems theory; and understanding the stair means more than a systematic collation; it is a sharpening of the senses and sensibilities. There have been many books that deal with stair construction and a few that deal with stair history. Some design pattern books and standard professional reference books include stair design criteria (usually quite removed from current knowledge). There is also an extensive literature that reveals something of the nature of the stair as one of the most dangerous manufactured objects. But there has been no book that scrutinizes steps and stairs vectorially in this way. In fact, even books and articles on stair history are largely limited to certain periods and certain countries; there is no adequate perspective view.

The evolution of the stair (and the ladder and the ramp) catalyzed human exploration and exploitation of architectural space. Spatial potentials demanded the invention of the stair, and the stair made possible quite new configurations and dispositions. Much later, the elevator (and the escalator) stimulated an explosion of vertical spatial experimentation and expansion into strata that lay well beyond the useful limits of stair access. But stairs were not superseded and rendered obsolete by these mechanical systems. Stairs have, of course, been retained as the backup system—a role that is no less important because it is supportive. The stair becomes the route of last resort where the potential for catastrophe demands a design that will always be safe and convenient. As often as not, these stairs are treated as a necessary nuisance and built to the minimum that codes will permit, with low levels of comfort, convenience, and ambient quality. And people are reluctant to use them, even for trips between adjoining floors. The resulting lack of activity sometimes transforms them into the ideal locus for criminal activity (Newman 1972). But the stair is an important architectural and architectonic element, still fascinating to designers.

In view of the continuing prevalence of stairs, their very ancient lineage, their glory as an architectural piece, and their dark side as places of great danger, it is remarkable that so little attention has been paid to them until recently; and there is still much to do. A lack of a system of design principles has obliged designers to make assumptions or to guess at what constitutes a stair that is safe, comfortable, and convenient. Even if all the rules of thumb, the design guidelines currently in use, and the extant safety and building codes are followed, the results may still be dangerously unsafe.

The Staircase attempts to remedy some of these omissions. These companion volumes describe investigations that probe many of the seemingly most important questions, and they suggest ways of considering others. This first volume (*History and Theories*) explores what can be deduced from the stair's past and the theories of design that were influential. The second volume (*Studies of Hazards, Falls, and Safer Design*) is concerned with the physical, physiological, and behavioral interaction between people and these Euclidian stair geometries. It is concerned with what we must expect from stairs and what we must do to control the stair's other, crueler nature, and it discusses quite a new idea, the soft stair, that may be able to reduce substantially the annual toll of deaths and serious injuries. Finally, the second volume sets out all the factors to consider when designing a stair. At our peril, we can continue to rely on archaic formulas, rules of thumb, a good eye, and experience; the published literature and the emerging research is too extensive and diffuse for most practitioners to keep up with. To put the information into practical form, the appendix to the second volume takes the designer through the affective factors and suggests ways to make design decisions. There are others—those concerned with code development or litigation, for example—who require a much more thorough discussion. For them, the rest of the second volume provides an introduction to current knowledge. These two volumes are an attempt to try to understand an element of architecture from many viewpoints: from the scientific, the cultural, the historical, the fanciful, the fearful, the behavioral, the biomechanical, the legal; the safe, the unsafe, the hard, the soft, stairs for comfort, stairs to astonish, minimal stairs, stunning stairs.

ACKNOWLEDGMENTS

Many people have helped me with the various studies that led to this book and the companion volume (*The Staircase: Studies of Hazards, Falls, and Safer Design*), and I thank them. Some have made extraordinary contributions. James Marston Fitch, my mentor at Columbia University, helped shape the holistic approach. His many questions constantly expanded the ideas and forced me to explore new views that I might well have bypassed. Paul Corcoran, another of my Columbia friends, introduced me to work physiology and gait analysis. Jack Fruin, whose published work on pedestrian planning is seminal, has given me many suggestions, particularly about the capacity of stairs to handle crowds. John Archea's work at the National Bureau of Standards greatly expanded our understanding of stair accidents and safety, and his contribution as a colleague on a number of stair research projects has been significant. Bob Brungraber has always generously given me help and advice on slip resistance. Jake Pauls's endeavors to encourage code recognition of his and our work have brought it into practical application. Ronald Lewcock, an old friend, generously read and criticized much of the manuscript and was particularly helpful with suggestions about the history of stairs. Craig Zimring provided critical suggestions about human behavior on stairs, and I have benefited from Harvey Cohen's collaboration on a project. David Lewis, Mark Coleman, Deborah Hyde, John Sanford, and many students made various research projects successful. Russell Waldon and my daughter, Nicolette, contributed extensively to the chapter on the law. The College of Architecture at Georgia Institute of Technology granted me a leave of absence for one quarter to finish the work, and the Georgia Tech Foundation generously contributed to the publication costs. Finally, Joan and Theresa Templer and Nicolette and Russell Waldon—my family—treated this project with tolerance and understanding.

The Staircase

1 Cordonata *in Santorini, the Cyclades, Greece*

INTRODUCTION

Paths are etched into the landscape by the passage of many feet. On a path that is steep and slippery, we climb by pushing our feet against a root or rock or a sod or into a rut. These footholds may develop into a pattern of rough steps, or what Rothery (1912) calls modified land stairs. A modified land stair or a ramp is not, by definition, geometrically uniform; typically they are irregular in layout—sometimes more like a series of stepping-stones than a flight of stairs.

To achieve some permanence and resistance to erosion by wind and rain, land stairs must be constructed of durable materials such as timber or stone. A spectacular example is the stone stair at Tai Shan, China. This so-called sky ladder ascends almost vertically up the side of the holy mountain to a shrine at the top.

Where the slope of the ground is gentle, the treads are often ramped to form the *cordonata*, or corded way. These timeless and ubiquitous ramps, modified land stairs, and *cordonate* form an essential part of the organic street pattern of hill towns like Santorini in the Cyclades and Moulay-Idriss in Morocco (figs. 1 and 2), responding to the vagaries of the topography. The rambling stepped streets of many Italian hill towns are modified land stairs. The patterns of their winding streets appear almost to have developed for the convenience of the pack donkey, as Le Corbusier

2 *Street stair, Moulay-Idriss, Morocco*

pointed out, as well as for human gait. At Sky City in New Mexico the Pueblo Indians cut a modified land stair into the steep, smooth rock face. This meanders down from the acropolis to the rock pool that contains the town's main water supply (fig. 3). They carved out the rock to form regular steps, shaped as if with "a geometry measured by the compass of a human step," as Gio Ponti (1960, 114) describes the proportional relationship needed between riser and tread. Even in this simplest form of constructed stair, humans show an understanding of the need for constant dimensions for ease of movement and for safety.

The modified land stair reaches its most developed form in the colonial cities of the Hellenistic world. At Priene in Sicily, for example (fig. 4), the stairs represent much more than an adjustment of the natural topography and access occasioned by coincidences of contours. Priene was planned before it was built; its stepped streets are called on to satisfy all the sophisticated functional criteria of the Greek polis. The mandatory Hippodamian gridiron is carefully set down on a rocky promontory. The streets are regular in width, straight, not too steep to climb, well drained, and they intersect at predetermined levels to form uniform *insulae* (blocks of houses).

Priene's display of geometric regularity provides a typological turning point—at least for the simple classification that follows. The shape of the stair begins to respond more to the deliberate decisions of the designer and the exigencies of the created structure to be served than to the natural contours of the land. Where the influence of topography becomes insignificant, then stair shape typically takes on a

3 Rock steps at Sky City, New Mexico

4 View of Priene, Hellenistic colonial town in Sicily

geometry quite different from the meanderings that are usual to modified land stairs.

For buildings, even the simplest and earliest ones, steps always display a regularity of layout and dimensional consistency. Even the sinuous curves of the baroque fantasies do not depart from these intentions. There are persuasive reasons for this Euclidean order. The stairs are easier to climb, safer to use, and easier to construct. However, even if no rational explanation can be offered, even if it is argued that these preferences are the consequences of archaic habits or human aesthetic instincts, the evidence that will be presented shows that stairs in simple societies, in ancient civilizations, and in complex modern technological cultures all adhere to these principles.

If I were a rich man/ If I were a wealthy man/ I wouldn't have to work hard/ I'd build a big tall house with rooms by the dozen right in the middle of the town/ A fine tin roof with real wooden floor below/ There would be one long staircase just going up, and one even longer coming down, and one more leading nowhere, just for show. (Fiddler on the Roof, lyrics by Sheldon Harnick)

Stairs serve many roles in addition to their prosthetic function. These roles may modify or even dominate completely the mundane purposes of safe, comfortable, and convenient ascent and descent. The stair has always been used to represent human spiritual aspirations and cosmography; to demonstrate secular power and authority, prestige and status; for aesthetic, architectural, and spatial manipulation; to make adjoining floors seem close, and the ascent a gentle transition; or to accentuate the separateness of spaces, with the staircase acting as a bridge. Stairs convey meaning and have personalities (figs. 5, 6, 7, and 8).

. . . agile, youthful, quick, flying, running steps; light, rustling steps; modest steps, shy steps; important steps; old, heavy, grave, slow, creeping steps; hard, fateful, fearful, frightening steps that make your heart pound; loving steps, thoughtful steps, murderous steps, terrifying steps, fugitive steps. (Ponti 1960, 117)

These innermost, almost archetypal responses to stairs carry with them cultural memories embedded in the psyche—part myth, part religious mystery, part dream, part fairy tale. Jacob dreams of "a ladder set up on earth, and the top of it reached to heaven: and behold the angels of God ascending and descending on it. And, behold, the Lord stood above it and said, I am the Lord God of Abraham thy father, and the God of Isaac" (Genesis 28:12–15). And the Persian mystic poet Jalalu'ddin Rumi also uses the metaphor of the ladder: "In the world there are invisible ladders, [leading] step by step up to the summit of heaven. There is a different ladder for every class, there is a different heaven for every [traveler's] way" (Rumi 1982, 114).

Ramon Lull (1235–1316), the medieval Christian mystic, depicts in the *Liber de Ascensu et Descensu Intellectus* (1512 edition) the Creation as a ladder of ascent and descent (fig. 9). Each step represents a level of creation, from senseless stone at the fundament, to flora, plants, brute animals, humans, the firmament, the angels, and, at the summit, God and the House of Wisdom (Yates 1966, 180) where the intellect may enter. For Perpetua, the Christian martyr slain in the public amphitheater in

5 Mouth of Hell, Bomarzo, Italy

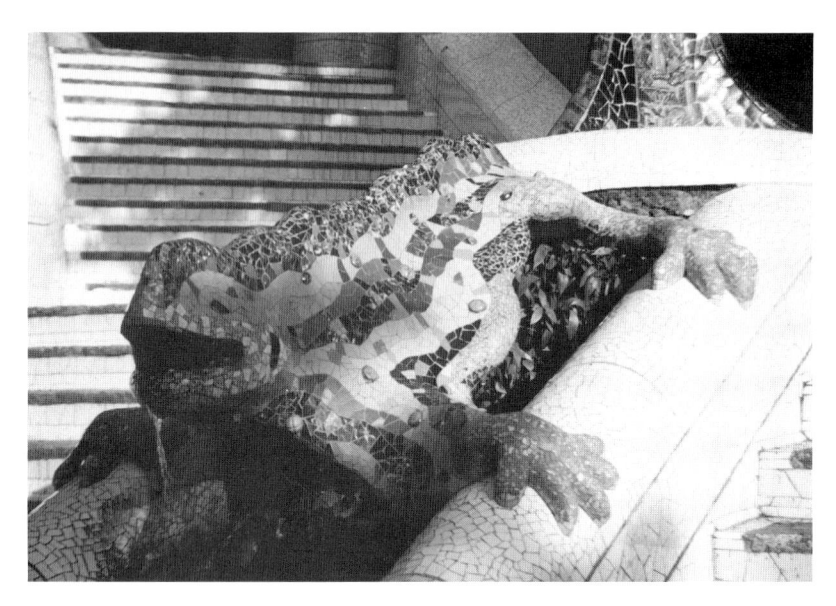

6 Gaudi's steps in the Guell park, Barcelona

Introduction

7 *Dubuffet's poetic steps, near Paris*

8 *Adalberto Libera's stair house, Capri*

Introduction

Carthage, the ladder (and the way) is painful and daunting: "She had dreamed that she was climbing a bronze ladder of tremendous height, bristling with daggers, swords, and spikes, reaching all the way to the heavens" (Pagels 1988, 34).

Sigmund Freud (1948, 355) dreams of stairs or sees stair dreams as sexual manifestations or representations, sexual fantasies transmuted into metaphorical icons: "We . . . began to turn our attention to the appearance of steps, staircases and ladders in dreams, and were soon in a position to show that staircases (and analogous things) were unquestionably symbols of copulation. It is not hard to discover the basis of the comparison: we come to the top in a series of rhythmical movements and with increasing breathlessness and then, with a few rapid leaps, we can get to the bottom again. Thus the rhythmical pattern of copulation is reproduced in going upstairs. Nor must we omit to bring in the evidence of linguistic usage. It shows us that 'mounting' is used as a direct equivalent for the sexual act."

For Henry Miller (1989, 286), mounting the staircase can be terrifying: "A sickish light spills down over the stairs from the grimy, frosted window. Everywhere the paint is peeling off. The stones are hollowed out, the banister creaks; a damp sweat oozes from the flagging and forms a pale, fuzzy aura pierced by the feeble red light at the head of the stairs. I mount the last flight, the turret, in a sweat and terror."

Ulrich Giersch (1983, 23–33) sees in Piranesi's Carceri series "steps in a labyrinthine intricacy down into the depths of the soul . . . into the dread unknown." Gaston Bachelard (1969, 57) reminds us that poetic

9 The Creation as a ladder of ascent and descent, from Ramon Lull's Liber de ascensu et descensu intellectus

daydreams reflect the "complex of memory and imagination." He quotes Rilke: "I never saw this strange dwelling again. Indeed, as I see it now, the way it appeared in my child's eye, it is not a building, but is quite dissolved and distributed inside me: here one room: there another, and here a bit of corridor which, however, does not connect the two rooms, but is conserved in me in fragmentary form. Thus the whole thing is scattered about inside me, the rooms, the stairs that descended with ceremonious slowness, others, narrow cages that mounted in a spiral movement, in the darkness of which we advanced like the blood in our veins."

This book discusses the history of stairs as cultural, political, architectural, functional, and aesthetic objects. The chapters are not arranged stylistically or chronologically but morphologically. I have not really escaped a chronological framework, nor have I tried to. By treating the stair in three chapters—straight flight, helical, and composite—I might appear to have opted for a formal framework, and this was probably true initially. However, the chapter sequence coincidentally reflects the chronology of the stair's major influences on architecture. The earliest stairs were straight flight; helical stairs had their heyday in the Middle Ages; and the great formal compositions were typical of the doglegs of the Renaissance, flowered in the baroque palaces, and have taken many forms since.

1.1 Sundial of the Jantar Mantar astronomical observatory, built in 1725 in Delhi by the maharajah of Jaipur

1 Straight Flight Stairs

A straight flight stair has a single flight or several flights linked by landings (fig. 1.1). The stair does not change direction. Conceptually at least, straight flight stairs range from climbing poles to ladders and companionways and from garret stairs to Michelangelo's masterpiece for the Laurentian Library. A straight flight ramp is in this sense a stair without risers.

1.1 Prototype: Climbing Poles and Ladders

The Dogon people of West Africa and villagers in Panama today still use one of the simplest and earliest types of stair, the climbing pole. It consists of a 6 to 12 inch (15–30 cm) diameter tree trunk that forks at the top. The fork rests against the upper part of the wall to be scaled. This steadies the pole, preventing it from revolving. Notches are cut into the trunk of the tree to form regularly spaced treads (the horizontal surface of a step) and risers (the upright face). This type of ladder has existed since earliest times. A similar type was used in the beehive neolithic houses at Khirokitia in Cyprus to reach the mezzanine floor level (Camesaca 1971, 16). Other variations lead to old Japanese and Lapp storehouses, and a pole carved into steps leads to the bowsprit (the spar that projects forward from the stem) of the U.S.S. *Constitution,* now moored at Boston.

1.2 Ladder at Sky City, New Mexico

Conventional ladders of twin supports with rungs may be as old as the climbing pole. The Modified Basket-Makers (c. 450–750) of Mesa Verde, Colorado, for example, entered their pit houses through the smoke hole in the roof by a ladder. Today at Taos and Acoma Pueblos in New Mexico, ladders are still used (fig. 1.2) to provide the external—and only—connection between the lower and upper floors and the roofs and the terraces of the adobe houses. This has been customary since the Developmental Pueblo Period (c. 750–1100) (Scully 1969, 19). Removal of the lower ladders discouraged unwanted intruders.

Houses in Tarong in the Philippines have a porch raised 5 or so feet (1.5 m) above the ground, with access by a bamboo ladder. At night or when no one is at home, the ladder is removed from the porch (Nydegger and Nydegger 1966), signaling that visitors are not welcome. Here the ladder serves as a formal entrance.

DEFENSIVE AND OFFENSIVE ROLES

Access was the primary purpose for some of the earliest stairs and ladders. However, it was often defense against humans and nature that made the separation between ground and dwelling necessary. Lifting the building only a few feet makes it safe from wild and domestic animals and even other people, and like all other defensive systems, it buys the defender a few moments of extra time.

In hot, humid climates or in marshy country, raising the hut enables it to enjoy a maximum of air circulation for cooling and to keep a relatively dry floor during torrential rains. In this respect, in the Papuan village of Barakau, where houses are built some 6 feet (2 m) above the water (fig. 1.3), the access ladders differ little

Straight Flight Stairs

conceptually from the three or four steps that separated Santa Maria della Pace from the mud and puddles of a seventeenth-century Roman street.

For military defense, the Spruce Tree house at Mesa Verde was built during the Grand Pueblo period (c. 1200) into a shallow cavern in the walls of a canyon. This dwelling and the two- and three-story adobe houses at Taos Pueblo (fig. 1.4) used external ladders that could easily be drawn up as protection from attack (Scully 1969, 23). The cliff dwellers at Mesa Verde also used a second, more permanent stair—a set of hand and foot holes carved into the cliff face leading from the canyon floor to the village. These holes are skillfully arranged. Climbers must start with the correct foot and hand or be left hanging in space just below the platform level of the village.

1.3 *Village of Barakau, Papua*

1.4 *Taos Pueblo, New Mexico*

Climbing Poles and Ladders

1.5 Defensive city wall, Morocco

There is a similar arrangement on the inside of defensive city walls in Morocco (fig. 1.5), where the hole ladder provides the only access to the battlements at the top.

For offense, siege ladders were basic military equipment (figs. 1.6 and 1.7). In fact, in the curiological writing of the Egyptians, the ladder symbolized the concept of siege (Foucault 1973, 111). Siege ladders could be decisive in an engagement. The crusaders failed in their first attempt to escalade the walls of Jerusalem for lack of enough ladders. There was a scarcity of wood and other materials in the surrounding country (Anderson 1970, 52).

In the Middle Ages, the ladder and the staircase were significant and carefully considered elements in the arsenal of offense and defense. Outside the walls of castles and fortified buildings, the stair's role was deliberately planned. The approach to the first floor was often by a stairway, which could be destroyed quickly before an attack. This was sometimes more like a ladder, of the type illustrated (fig. 1.8) in the Bayeux Tapestry, leading to a Norman motte wooden tower. If the stair was permanent, it usually led to a drawbridge or a removable wooden platform. Many of these outside staircases were covered by a roof, often made of wood that could be burned or otherwise removed before attack, ensuring that the enemy would have no cover for the assault. Sometimes the approach stairway to the first floor was built parallel to the outer walls, clinging to the sides of the castle. An enemy brave—or foolish—enough to attempt to climb an exposed staircase of this type would have to run the gauntlet of arrows, stones, hot oil, and boiling water showered down from the embrasures above.

Straight Flight Stairs

1.6 Siege ladder, from Eugène-Emmanuel Viollet-le-Duc, Dictionnaire Raisonné du Mobilier Français, 1874

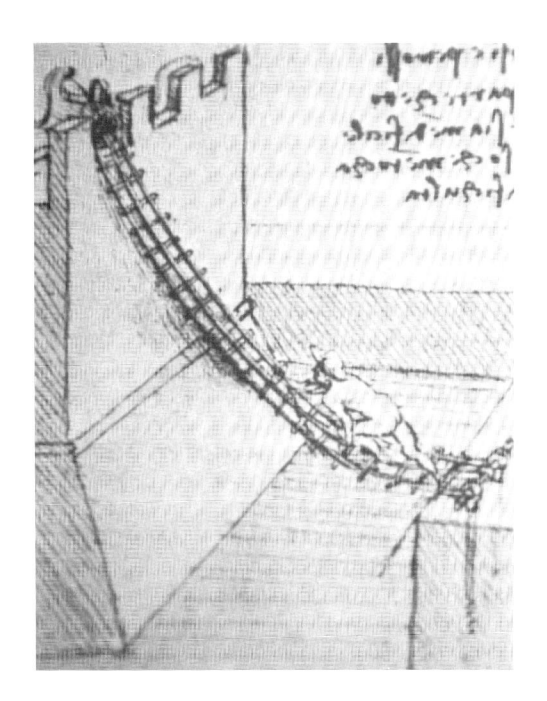

1.7 Siege ladder, sketch by Leonardo da Vinci

1.8 Norman motte in the Bayeux tapestry

Climbing Poles and Ladders

FROM RUNGS TO TREADS

To prevent the supports from bending too much and to make climbing easier, the ladder and climbing pole must be set against the wall at a fairly steep angle. People climbing must hold on to the sides or rungs to keep their balance; no one can climb a ladder while carrying much.

The companionway of a ship is also fairly steep usually, but it has treads rather than rungs. Weight can be spread over a much larger area of the feet rather than concentrated on the surface of the small rungs of a ladder. Therefore, it is possible to carry much heavier loads. However, the companionway is functionally usable only because it has handrails to steady the climber. The body's balance still cannot be maintained easily over such a narrow base without this support. Usually we descend a companionway by backing down; there is too little exposed tread to support more than the back of the foot if we try to descend facing forward. Decreasing the angle of the gradient of the companionway increases the available tread. When the gradient angle is made sufficiently small and the treads made sufficiently large to stand on them while descending facing forward, we have a stair rather than a companionway.

The Metropolitan Museum of Art, New York, contains a painted wooden model (fig. 1.9) of a granary from the Eleventh Dynasty (c. 2000 B.C.) Egyptian tomb of Meke-Re at Thebes. The model has a stair

1.9 Egyptian Eleventh Dynasty model of a granary

that is something between a companionway and a ramp (but it is not a true stepped ramp). Here, if we can believe the accuracy of the model, the granary workers are carrying sacks of grain up a stair with very steeply ramping treads and small risers and, despite the steepness, no handrail. The ramping of the treads reduces the height of the risers. The treads are made sufficiently large to provide quite a large base for balance purposes. The idea is ingenious, but the ramped treads look as if they may be dangerously steep for descent, particularly if they become slippery from grain, dust, or moisture.

1.2 ARCHETYPE 1: THE ARCHITECTURE OF THE INTERNAL STAIR

Of the three straight flight examples dealt with so far—the ladder, the companionway, and the climbing pole—only the companionway usually has a fixed location. Lightness and ease of transport are of little advantage for stairs whose position in relation to the building will be unchanging (except perhaps for defense). Furthermore, fixed stairs are inherently more stable. They are safer and much easier to climb, and climbers can carry much greater loads. Fixed stairs have been built wherever, and whenever, these attributes begin to outweigh the advantages of the ladder.

The straight flight stair is the archetypal stair. It has been found in excavations of buildings in Egypt (fig. 1.10) and Mesopotamia dating from earliest times and remains in common use today (fig. 1.11). It is the simplest stair in layout; the builder does not require a sophisticated structural knowledge. It is relatively easy to accommodate within a building.

For access to upper floors, the straight flight stair's advantages of simplicity and directness are offset by the need for the stair to rise comparatively steeply. Flights with gentle slopes are longer, with the point of access located often inconveniently far from the point of egress, and stairs with small gradients occupy more floor space.

The zigzag or double-riser stair is a curious and venerable device invented to halve the usual horizontal run. Notre Dame in Paris uses one to gain access to the roof (fig. 1.12), and there is one in Mont-Saint-Michel, the medieval abbey-island off the coast of Brittany in France. Its lack of popularity may be a reaction to its visual steepness and because one must always step off with the correct foot. If one must pause on the stair, one stops with one foot above the other, and passing is impossible. Furthermore, it is more physically taxing than more conventional designs. Nevertheless, it still reappears from time to time. Sir Edwin Lutyens's remodeling of Lindisfarne Castle (fig. 1.13) on Holy Island, Northumberland, England, uses a modern version of the zigzag stair, and a commercially manufactured version, the Lepeyre industrial stair (fig. 1.14), is something between a ladder and a true stair. It has the same double-height risers set on either side of a center spine, and although its pitch is 70 degrees or so, those descending can face forward.

1.3 THE STAIR AS A PASSAGE

ARCHITECTONIC DEVICES

In its simplest form—in the traditional Mediterranean house, for example—the straight flight stair, little more than a sloping passage with walls on both sides, is

The Architecture of the Internal Stair

1.10 *Stair at the temple of Hatshepsut, Thebes, 1520 B.C.*

1.11 *Pulpit stair, Le Corbusier's chapel at Ronchamp, France*

1.12 *Double-riser stair, Notre Dame, Paris*

Straight Flight Stairs

The Stair as a Passage

1.13 Double-riser stair, Lindisfarne, England, by Sir Edwin Lutyens

1.14 Lepeyre industrial double-riser stair

Straight Flight Stairs

usually tucked away out of sight. It is not exploited as an architectonic device to connect the two floors visually. One experiences the stair passage with a tunnellike view of the space to which the stair leads. Traditionally, these stairs frequently lacked handrails and were precipitous and ill lit even in patrician houses. Vitruvius (1926, 185), the author of the first known text on architecture (46–30 B.C.), says little about domestic stairs, but he draws special attention to the need for stair illumination: "Not only in dining rooms and other rooms for general use are windows very necessary, but also in passages, level or inclined, and on stairs; for people carrying burdens too often meet and run against each other in such places."

Even these ramping tunnels, however, offer the possibility of spatial experience. They can prepare us gradually for the space that we are approaching by controlling our view of what is to come. We can experience them as architectural elements as we pass along them. And as architectural components that are part of a whole composition, they can be viewed from various places as we approach them, pass them, or simply observe them from afar.

GRADIENT, GAIT, AND IMPORTANCE

Stairs engage the user's motions and their senses to a remarkable degree—perhaps more so than any other architectural element. The enclosing balustrades (or walls) of the flight control the stair user's movement through the space, and the dimensions of the risers and treads strictly govern the cadence of gait.

Strictly utilitarian stairs, both ancient and modern, with comparatively small treads and large risers, act to propel us along the stair at a comparatively brisk and businesslike pace. Stairs with larger treads and smaller risers encourage us to employ a more leisurely gait, permitting us to linger longer on the stair, to pass more slowly, and to spend more time in sensing the nature of the stair's setting and its spatial and decorative qualities. Consequently, gentler slopes appear early where ceremonial and monumental stairs are desired. They are the design of choice where a visually interesting architectural scene warrants a slower pace or where the stair itself is expected to be impressive.

The chateau of Vaux-le-Vicomte outside Paris has an interesting variable-tread stair. The chateau was designed by Louis Le Vau with André Le Nôtre, the garden designer, for Nicolas Fouquet, lord high treasurer of France. The work was completed in 1661 just before Fouquet's trial and imprisonment for financial malpractice. An axial path from the palace terminates at a great wall and water feature overlooking the canal set at right angles to the main axis. On either side of the wall, ramps and, finally, steps lead down to the large space framed by the canal and the wall. The treads of these two flights of steps increase in size from top to bottom (fig. 1.15). The tread at the top is 13.75 inches (35 cm) deep, and the bottom tread is 20.5 inches (52 cm); and the increase in size is consistent. Each riser, on the other hand, is about 6.33 inches (16 cm) high. There is no record to tell us why Le Nôtre made the stair in this way; however, the effect of the device is to cause one to adopt an increasingly stately pace as one enters what was one of the main entertainment areas of the garden. (One might expect this to be a hazardous gambit, and dimensional inconsistency is discussed in the companion volume of *The Staircase*.)

The Stair as a Passage

1.15 Variable-tread steps at Vaux-le-Vicomte, outside Paris

With a growing understanding of the significance of the mathematical relationship of riser to tread came the realization that certain combinations are more appropriate for certain circumstances. The vomitory stairs of the Roman Colosseum are steep by any standards but are appropriate in principle for their functional purpose of moving crowds of people quickly. But it is difficult to visualize the ladies of, say, the court of Louis XIV, with their gowns formed from layers of petticoats, skirts, and overskirts, trying to descend these stairs. The gentle slope of the great baroque stairs, by comparison, was appropriate for the stately dress, manners, and movement of those ceremonious times.

Sequential Spatial Manipulation

As an architectural space, the stair was gradually embellished. The corridor box of the flight was articulated and expanded spatially with painted scenes and decorative motifs. Later, arches, columns, vaults, and molding were used to break up the passage into a series of spaces to be experienced sequentially. (The stair, by nature, is a sequential experience.)

Additional emphasis on the space was achieved by pushing out the walls and the ceiling to make much grander spaces. Monumentality was created by several other devices: the tread was enlarged, the riser decreased, an excessively generous width of staircase was provided, the size and importance of the (apparent or real) handrail was exaggerated, and an even more dramatic scale magnification was created using false perspective. Pevsner (1968, 185) describes Bernini's Scala Regia in the Vatican (1663–1666) (figs. 1.16 and 1.17): "It [the site] is long, comparatively narrow,

1.16 Falda's view of the Scala Regia in the Vatican

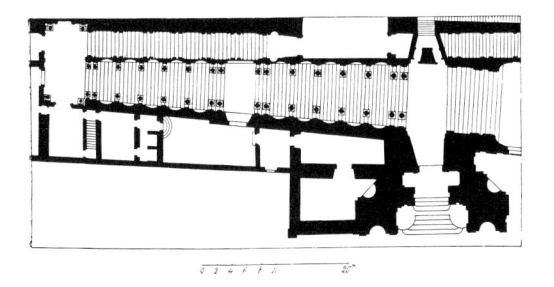

1.17 Plan and section of the Scala Regia

The Stair as a Passage

and has irregularly converging walls. Bernini turned all this to advantage by means of an ingenious tunnel-vaulted colonnade of diminishing size. The principle is that of vistas of the Baroque stage. . . . Light is another means of dramatizing the ascent up the Royal Staircase. On the first landing half-way up it falls from the left, on the second in the far distance a window faces the staircase and dissolves the contours of the room." Bernini's false perspective is a masterly solution to a perplexing problem: he had to fit his new staircase into a wedge-shaped space, using the existing wall on the one side and the upper landing and flight on the other side.

THE STAIR OPEN ON ONE SIDE—VISUAL FOCUS AND SPATIAL CONNECTOR

All of these devices contribute to the experience of the stair itself as a step-by-step experience. Partially or completely removing the enclosing walls of the stair increases the sense of spatial connection between stair and space and between adjoining floors. One may sense the space of the stair enclosure and simultaneously the changing views of the outside space through which the stair enclosure passes. Charles Moore carefully exploits this "moving focus," as he calls it, in the Stern house, Woodbridge, Connecticut (1970). He exaggerates the spatial tension by using stairs and long galleries on interlocking axes: "The main spaces of the house are stretched out in passages inhabited by people moving. If you see a corridor (and a stair) as non-room, as wasted space, then this house is wasteful. If you see it as a room stretched, an empty stage for moving as well as resting, then here are rich chances for improvisation" (Moore, Allen, and Lyndon 1974, 102).

Simpler versions of these devices date from the earliest times. In the House of the Surgeon (third and fourth century B.C.) at Pompeii, for example, there are two stairs completely open on one side. One of these stairs connects the colonnades to the rooms over the rear of the house, and the other leads from a shop to rooms over the front of the house. The diagonal line of the exposed stair, the only major sloping plane in a composition of horizontal and vertical planes, forms a strong visual connection between the two floors. One experiences the volume through which the stair passes as a more important visual event than that of the stair itself. In these early examples, however, the character of the stair is given little prominence in comparison to that of the space through which it travels.

The grand stair at Hardwick Hall (1590–1596), on the other hand, is given extraordinary emphasis for its time and place. The theatrical baroque stairs of Europe took root in England in a much more modest form and much later; the typical sixteenth-century grand stair in England is quite small. It has a square or rectangular plan with an open well. Hardwick is a remarkable exception and reflects the willful autocracy of Elizabeth (Bess of Hardwick), Countess of Shrewsbury. Robert Smythson designed a matching pair of open-well square staircases for the symmetrical plan, but the countess had other ideas. The grand stone stair that she built (fig. 1.18) is a series of straight flights, connected by landings, and culminating in a sweeping segment of a helical stair. The main part of the stair stretches for much of the south wing of the house up to the state rooms that are, curiously, placed at the top of the house. The stair itself is without decoration

Straight Flight Stairs

1.18 Grand stair, Hardwick Hall, England

1.19 Snyderman house, Fort Wayne, Indiana, by Michael Graves

The Stair as a Passage

and without even a balustrade, but the open volume within which it lies is hung with tapestries. The stair, unique in Elizabethan England, was the most spatially dramatic of its time.

The exterior of the Snyderman house, Fort Wayne, Indiana (1972), by Michael Graves uses a quite modest stair to tie the upper two floors together visually, spatially, and functionally (fig. 1.19). We are treated to a highly complex experience. On one side, we are suspended in space, cantilevered beyond the building, held between the sky and the ground. On the other side, we pass by the outer framework of the building and through it see changing views of a two-story terrace that is open to the sky and encloses and exposes other parts of a structural grid. On the facade, the strong diagonal line of the stair is used as a foil for the orthogonal grid, in a way that was favored by medieval builders. The eye follows the diagonal to explore the formal complexities of the upper part of the terrace.

THE DIAGONAL ACCENT

Buildings are composed predominantly from horizontal and vertical planes, and either may be exaggerated by the designer. Medieval Gothic religious buildings emphasize the vertical leading the eye to the heavens as a contrast to the earth plane of the flesh. Frank Lloyd Wright preferred horizontal planes in harmony with and reflecting nature and the earth. The diagonal line or plane is comparatively unusual in the major massing of building components except as a roof cap, and architects have always sought to understand and to tame the vigorous, unruly heresy that the diagonal demonstrates within comfortable orthogonal schema. The nature of the diagonal is a forceful dynamic movement that may threaten the tranquility of the usual order and orientation. It has shock value; it surprises us and rivets our attention and is often used deliberately for this reason, as in Richard Rogers and Renzo Piano's Pompidou Center in Paris (fig. 1.20).

The stair is diagonal by nature, strengthening the connotations of movement implicit in the sequence of risers and treads. The medieval world enjoyed and exploited the diagonal of stairs; it was a natural element to balance asymmetrical arrangements (fig. 1.21). Goethe was, in fact, convinced that the Pisa campanile was intended to lean as an accent to contrast with the hundreds of vertical towers of the city. The early Renaissance architects, on the other hand, tended to put stairs into their own compartments to prevent them from violating the even balance of horizontal and vertical components of the rest of their compositions. The mannerist-modernist Renaissance of the twentieth century found new pleasure in the force of the diagonal line. Stairs and ramps have been given a new emphasis in compositions (fig. 1.22) in contrast to their decrease in importance in the general movement system. Elevators and escalators, the technological successors to stairs, have become the new force within great internal spaces, and the diagonal of the moving stair inserts a powerful visual movement into otherwise fairly static compositions.

RAMPS

In many ways, ramps are much like stairs as architectural devices, but because their slope is small, they occupy more space.

Straight Flight Stairs

1.20 Pompidou center, Paris

1.21 Palazzo dei Papi, Orvieto

The Stair as a Passage

1.22 Mill owners' association building, Ahmadabad, India, by Le Corbusier

They do not, however, present the sequence of visual (and mobility) barriers of a flight of steps, so they are much safer. We need not concentrate so carefully on maintaining balance and foot placement. This releases our attention to respond more freely to our surroundings; we can use the ramp with almost as much inattention as we would give to a level walkway. In this way, compared to stairs, ramps are less of a spatial barrier between spaces and adjoining levels. In their search for devices to reduce spatial containment and to accentuate the visual flow of space, the architects of the International Style quickly discovered this attribute of ramps, probably inspired by Le Corbusier's little masterpiece, Villa Savoye at Poissy (fig. 1.23). The ramp became one of the typical stylistic trappings of the movement and certainly part of the vocabulary of spatial freedom (fig. 1.24).

Straight Flight Stairs

1.23 Villa Savoye, Poissy, by Le Corbusier

1.24 Brasília, by Oscar Niemeyer

The Stair as a Passage

Stairs Open on Both Sides—Staircases as Objects in Space

Straight flight stairs that are open on both sides are hardly new; there are innumerable examples in the ancient world. Florence and Naples offer particularly splendid examples. For the Laurentian Library (1557) in Florence, Michelangelo utilized an array of architectural devices (fig. 1.25). The stair is open on both sides but set into its own enclosed envelope, a stair hall—that great internal space that the German and Viennese baroque masters called the *Treppenhaus*.

Medieval architects delighted in the invigorating dynamics of diagonal (and curved) lines and seldom hid stairs from view except for reasons of defense or economy. Often they were exposed on the outside and their climbing shapes expressed. The architectural theories of the early Renaissance, however, required the freedom of movement manifest in a staircase to be rendered as statically as possible. The diagonal forms of the stair interfered with the cubic perfection aimed for in the rest of the spaces. The solution was to enclose the stair; hence the *Treppenhaus*. The Renaissance architects nevertheless reveled in the potential of the stair as architecture and soon chose to increase its importance by increasing the size of the stair and its volumetric container. The culmination of this trend is manifest in the magnificent baroque stairs of Italy, Austria, and Germany.

In the Laurentian Library stair, Michelangelo made the stair into a freestanding object attached to the wall on only one side. The central main flight is accompanied for part of the way by secondary flights on each side. The function of these companion flights was probably to accentuate the monumentality of the stair and, it has been suggested, to provide places where uniformed footmen carrying lanterns could stand for formal occasions. The two outer flights are flanked by giant steps twice the size of the others that emphasize the monumental scale of the stair.

The steps of the stair appear to flow out of the doorway to the upper level as if the whole stair is a stream. And, as we shall see, garden and fountain design strongly influenced Renaissance (and baroque and mannerist) architects. The flow of the steps is broken into three cascades by landings that form pools, and the extremities of the steps form volutes as if these are eddies in the current. The marble stair fountain then floats in a clay tile floor as if in a pool bridged by the top flight. This stair is a sculptured object set within a carefully modulated enclosing box that completely contains the space, frames the stair, and does little to distract the user from the dramatic volumetric experience of the rising planes of the staircase. In fact, the walls of the stair hall are treated as if they are the outside walls of a building. This staircase helped to expand the conception of the stair as a decorative element. The baroque architects, who developed the staircase into such a dramatic component of architecture, reveled in the sensuous linear qualities of the curves and volutes of the treads of this Medici stair.

Curved treads and risers were used for the *crepidoma* (the stepped base) of classical Greek and Roman tholoi, such as Theodorus of Phocea's delicate little temple at Delphi (390 B.C.) (fig. 1.26). Curved steps have also been used to form theater seating since earliest times. Until late in the Renaissance, however, curved steps were seldom used. Michelangelo boldly freed the

Straight Flight Stairs

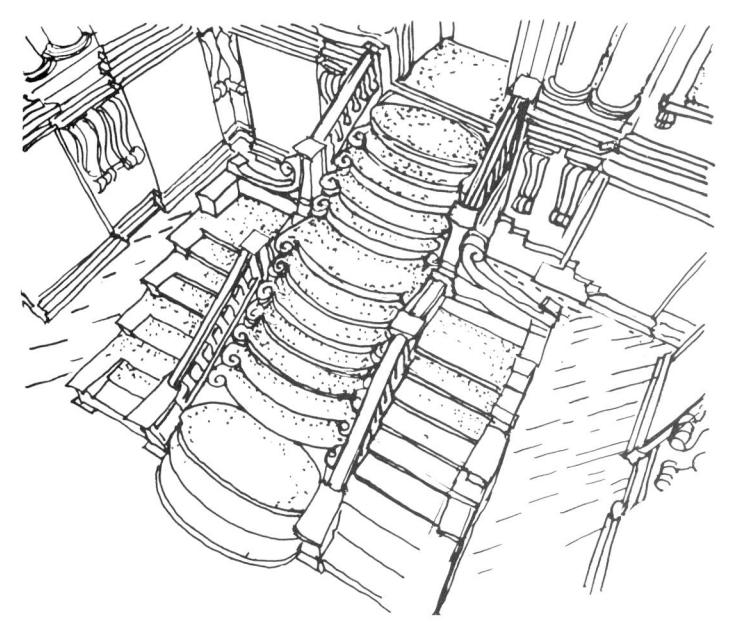

1.25 Michelangelo's stair for the Laurentian Library, Florence

1.26 Tholos at Delphi

The Stair as a Passage

straight flight stair from its tradition of rectangular treads. Leonardo probably did the same for the helical stair some years earlier at Blois in France. Michelangelo (like Leonardo) treats the stair as sculpture to be used rather than as a useful device to be adorned with sculpture. The treads are not decorated but are decorative. The nosing line bows outward but then changes direction abruptly to form the voluted ends, a decorative artifice that strengthens the visual comparisons of the stair to a motionless waterfall or a flow of molten lava.

Particularly for external stairs, the use of this abrupt change of direction and the curvilinear line of the nosing adumbrates the visual excitement and richness of the Scala di Spagna (and must have characterized the Ripetta steps to Rome's now-defunct harbor on the Tiber).

THE STAIR UNCAGED

It was Rocco Lurago and Michelangelo's pupil Alessi (1500–1573) and their followers in Genoa who freed the straight flight stair from its Renaissance cage, where it was tightly enclosed within its own stair hall. These Genoese architects allowed the stair to be seen at a distance and enjoyed as both a decorative object in space and a means for visually connecting spaces. The vertical volume of the stairs in Lurago's Palazzo Municipale (1564) and Bartolomeo Bianco's Palazzo dell'Università (1623) (fig. 1.27) and Palazzo Marcello-Durazzo (1619) (fig. 1.28) are still contained in their own halls, in this case the entrance lobbies. The vertical volume of the shady lobbies, however, is fused with and contrasted to that of the sunny courtyards above, into which they lead. The staircases lead the eye upward to the volumes of the courtyards. The flow of space from the lobby to the courtyard, from the one level to the other, takes place but without destroying the integrity of the two volumes. It is an ingenious solution to a difficult Renaissance design conundrum.

1.4 ARCHETYPE 2: THE ARCHITECTURE OF THE EXTERNAL STAIR

STAIRS OF THE GODS

Sigfried Giedion (1964) equates the rise of the first high civilizations with the rise of the vertical as the preferred direction. Rapoport (1969, 8) suggests that the concept of verticality emerges quite late in civilizations. "There is evidence," he says, "that some present day 'stone-age' civilizations such as the Eskimos show a lack of differentiation conceptually between directions and also lack a preferred direction in their art." And Norberg-Schulz (1971, 21) writes that "the vertical, therefore, has always been considered the sacred dimension of space. It represents a 'path' toward a reality which may be 'higher' or 'lower' than daily life, a reality which conquers gravity, that is, earthly existence, or succumbs to it."

In the elaborate cosmography of some preindustrial societies such as that of the Nias islanders of Indonesia, the village is, in the minds of the people, synonymous with the cosmos and is almost invariably sited on hilltops. Indeed, the Nias word for village (*bahmua*) means "sky" or "world." Access to the village of Bawamataluo in South Nias is gained by climbing an absolutely straight and regular flight of stairs and terraces set to line up symmetrically with the rigidly axial main street. This formal approach stair ties the village to the

1.27 Palazzo dell'Università, Genoa

1.28 Palazzo Marcello-Durazzo, Genoa

The Architecture of the External Stair

rest of the world: "The stone stairways leading to the village are carved with images of crocodiles, lizards, and other symbols of the lower world, some of them devouring fish, dogs and other animals. This end of the village (JOU) is regarded as 'downstream' which is synonymous with death, commoners, aquatic animals, 'west' and 'north'" (Fraser 1968, 38).

Heaven is "up," hell is "below"; we climb the "ladder of success," take the "stairway to the stars"; our spirits are "up," we are in the "depths of despair." Since so many of this world's gods have chosen to live on and instruct from the mountaintops of Olympus, Sinai, Ida, Fuji, and the Himalayas, it is not surprising that people should have built ziggurat "mountains" (fig. 1.29) to lure the gods into human settlements or to bring people closer to the gods (the word *ziggurat* stems from the Assyrian *Ziqqurat,* "summit" or "mountaintop").

The builders of these homes for the gods were careful to ensure that these places are approached with suitable humility. Three flights of stairs led to the portico of the lowest tier of the ziggurat of Urnammu at Ur. They formed an approach and an entrance as demandingly obvious and unambiguous as the great ramps of the temples of Mentuhotep and Queen Hatshepsut at Deir el Bahari (fig. 1.30). One is at once dwarfed by the scale of the approach, mercilessly exposed, and in essence reminded of one's insignificance in the cosmos and in the realms of the gods, the king, and the powerful. The monumental staircase is born.

Similar characteristics are evident in the pre-Columbian building complexes, where often the ziggurat itself seems like a pyra-

1.29 Tower of Babel, by Livius Creyl, 1670

1.30 *Funerary temples at Deir el Bahari*

midal stair for giants. The so-called Castillo of Chichén Itzá in Mexico is like this. Superimposed on each of the four stepped sides is the other stair, the one for human feet (fig. 1.31). From all sides, the upper temple is seen to stand at the top of a compellingly direct monumental flight of 365 steps, unrelieved by any feature to distract the eye from the focus at the top.

The route to the top of the pre-Columbian pyramid is not made comfortable physically or psychically, for the Mayan-Toltec builders, who at the Nunnery at Uxmal provide quite short and gentle approach stairs, present the climber of the pyramid of the Magician at Uxmal (fig. 1.32) with a towering flight of nearly 100 steps rising from the ground at about 60 degrees, without a landing of any sort on which to pause and catch one's breath and balance. The stair is a ceremonial way, acting like a stadium in reverse, where the plebs can see the priestly processions and rituals acted out on the stairs (figs. 1.31, 1.32, and 1.33). The act of climbing becomes ritually significant.

The symbolic and ritual use of stairs in pre-Columbian architecture extends far beyond that displayed by individual buildings. Machu Picchu, Peru (fig. 1.34), Monte Albán, Mexico (fig. 1.35), and Teotihuacan, Mexico, are all immense urban compositions of stairs—small steps to walk on, huge steps made up of smaller steps, stepped platforms, steps on steps—all leading the eye and the processional paths upward. At Machu Picchu (fig. 1.36), this culminates at the extraordinary informal arrangement of steps carved into the Intihuatana, the altar or sundial. To this the sun god was tied in an annual ceremony at the winter solstice to prevent him from slipping away farther, and to encourage the deity to return, bringing summer.

The stupas and pagodas of the Orient are also imbued with a cosmological ordering. Each is a temple-mountain, an axis mundi, a sacred center in which the infernal, terrestrial, and celestial worlds are represented. Their layers or stories represent the celestial regions that must be traversed in the inner journey upward through the heavens toward enlightenment, perfection, and ultimate reunification with the Creator. They seek to make visible the world of truth beyond our world of the senses, of appearances. This is an architecture of sacred images.

The name of the great ninth-century Barabudur in Indonesia (fig. 1.37) means "The Mountain of the Accumulation of Virtue in the Ten Stages of the Bodhisattva." Its hemispheric silhouette recalls a stupa, an early Buddhist shrine. *Stupa,* from Sanskrit, means the "crown of the head." So Barabudur represents the hemisphere of the heavens as a replica of the cosmos. It also signifies the sacred mountain, and the forehead from where the meditational inner journey starts.

The structure consists of five steps or stories supporting three circular terraces. On each of the four sides, flights of steps lead to the top. According to Buddhist cosmology, the creation is divided into three parts. The highest sphere is formless, pure spirit, Ultimate Reality. Below this is the sphere of forms that lies above our own phenomenal world. At Barabudur, the pilgrim ascends the five terraces representing the world of forms, and the higher and higher spheres of spiritual life (Kempers 1959, 44).

Straight Flight Stairs

1.31 Castillo, Chichén Itzá, Mexico

1.32 Pyramid of the Magician, Uxmal, Mexico

The Architecture of the External Stair

1.33 Castillo, Chichén Itzá

1.34 Machu Picchu, Peru

Straight Flight Stairs

1.35 Monte Albán, Mexico

1.36 The Intihuatana, Machu Picchu

The Architecture of the External Stair

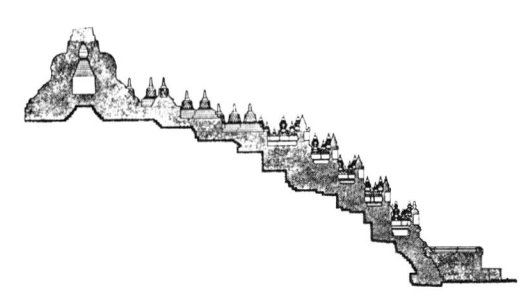

1.37 The Barabudur, Indonesia, ground plan and vertical section, reprinted from Kempers (1959), 42–43

SCALE, SCALA, AND PROPORTION

It did not take long for the designers of buildings to realize that stairs can also convey an idea of relative size. Stairs are one of the few devices in architecture that can indicate to the observers just how big a building, or part of a building, is compared to other parts or to its surroundings. Even the word we use for this purpose, *scale,* shares the same etymological roots as the Latin word for stairs (*scala*).

Stairs can perform this scaler function uniquely well because the dimensions of risers and treads must conform to the comfortable limits imposed by human gait. François Blondel's prescription of these limits in his famous formula (twice the riser plus the tread equals a constant) is close to the truth. As he demonstrated through a rather neat empirical study in about 1672, we cannot comfortably use risers that are greater than about 8½ inches (22 cm) or a tread that is much less than about 9 inches (23 cm). Treads can be quite large, but risers cannot be. Steps cannot vary much in size. Therefore we can often deduce building size from steps.

Many size clues are available to us, and our stereoscopic vision gives us much of our depth perception and distance judgment through the laws of perspective. However, the farther an object is from us and the larger it is, the more difficult it is for us to estimate its size accurately. Other clues are available. We can, for example, compare the building to familiar objects—people, cars, and so on. Absent any of these clues, steps and stairs are the most reliable scaler element for buildings. The stairs in front of a row of New York brownstones (fig. 1.38) tell us more about the size of the buildings than any other feature except perhaps the curb.

Straight Flight Stairs

1.38 *Brownstones, New York City*

1.39 *Access steps to the Parthenon, Athens*

Windows, doors, and columns cannot be trusted to help us estimate size, for their dimensions can be manipulated. Saint Peter's in Rome is the most frequently quoted example of this. From a distance, all of its parts appear to be quite usual in size. As we approach, we begin to realize that its doors, windows, and columns are much greater than we expected.

Even apparent stair size cannot always be trusted. If the stairs that we examine are not intended for human gait, they may convey a false message. For example, one may be persuaded that the steps of the Parthenon's crepidoma are for humans. They are not; they are an intentional scaler manipulation; the treads are 28 inches (71 cm) and the risers 20 inches (53 cm)—three times life size. By the time we find the true human stairs on the side of the temple, we have discovered that once again we have been deceived, and this most human of buildings is much larger than we expected (fig. 1.39).

Platforms and Podia

Elevating temples, public, and religious buildings, symbolically and actually, above the earth plane and humanity was customary. Even the meanest temple in the classical world is raised on a crepidoma, and the number and size of these steps has a symbolic and ritualistic significance. Vitruvius (1926, 88) writes that "the steps in front [of temples] must be arranged so that there shall always be an odd number of them; for thus the right foot with which one mounts the first step will also be the first to reach the level of the temple itself." Rothery (1912, 16) suggests that this has to do with the ritual of approaching the altar with right hand uplifted in submission and supplication. However, Francesco di Giorgio Martini, writing in his *Treatise on Architecture, Civil and Military* (1482), evidently favors the other foot, for he states that "stairs should always be stepped on with the left foot." He offers no reason for his statement.

The steps leading to the choir in some medieval churches are also symbolically significant. They represent the mountain from which Christ delivered his sermon (or perhaps Calvary or the hill of Golgotha), and the frequent use of fifteen steps to the choir represents the fifteen virtues. The three steps to the altar symbolize virtuous faith, love, and hope (Mielke 1966, 150). Pilgrimage steps were constructed so that the pilgrim will strive, step by step, to attain a worthy goal and shed worldly concerns.

Straight Flight Stairs

The Scala Sancti (or Scala Pilati, as it was sometimes called), which leads to the Sancta Sanctorum, the Palatine chapel of the popes in Rome in the old Lateran Palace, is the best known of these. Medieval tradition has it that these were the steps Jesus climbed several times on the day he was sentenced to death in Jerusalem. Pilgrims climb the twenty-eight marble steps on their knees offering a prayer on each. On a grander scale is the pilgrimage church of Bom Jesus do Monte at Braga, Portugal (fig. 1.40), where thousands of pilgrims ascend the monumental baroque stair at Whitsun following a Way of the Cross.

In medieval Europe, it was not unusual to find great cathedrals resting on a substantial podium, evidently the original situation of Notre Dame in Paris, according to Victor Hugo (1947, 98):

That facade, as we now see it, has lost . . . the flight of eleven steps which raised it above the level of the ground. . . . Time, by a slow and irresistible progress raising the level of the city, occasioned the removal of the steps; but if this rising tide of the pavement of Paris has swallowed up, one after another, those eleven steps which added to the majestic height of the edifice, Time has given to the church more perhaps than it has taken away; for it is Time that has imparted to the facade that sombre hue of antiquity which makes the old age of buildings the period of their greatest beauty.

During the Renaissance and baroque, under the philosophical influence of the renewed interest in the ideas of the classical world, a church that did not sit atop its own stepped podium looked mean, visually uncomfortable, and somehow lacking

1.40 *Bom Jesus do Monte, Braga, Portugal*

The Architecture of the External Stair

in dignity—at least to the architects. Consider the quite preposterous (for the time) attempt to remedy such an omission, made by Leonardo da Vinci. Leonardo proposed to lift the baptistery of Florence Cathedral up into the air and then to mount it onto a podium. Vasari (1946, 188) writes that "among Leonardo's models and drawings is one by means of which he sought to prove to the ruling citizens of Florence, many of them men of great discernment, that the church of San Giovanni [presumably San Giovanni Battista, the baptistery to the cathedral] could be raised and mounted upon a flight of steps without injury to the building. He was so persuasive that it seemed feasible while he spoke, although every one of his hearers, when he was gone, could see for himself, that such a thing was impossible." Outrageous as the idea might have seemed, the medieval baptistery (c. 1150) does seem to squat rather heavily on the ground.

A podium serves several purposes. It lifts the building up from its surroundings, making it appear more important and separating it from its more mundane neighbors. The building is placed on a pedestal and given a base. Santa Maria Maggiore in Rome, for example, is raised on a large stepped decorative base. The podium's horizon planes are formed into a pattern of sweeping curvilinear elements that underline the profusion of textural and formal virtuosity of the facade above (figs. 1.41 and 1.42). For the faithful, climbing the

1.41 *Santa Maria Maggiore, Rome*

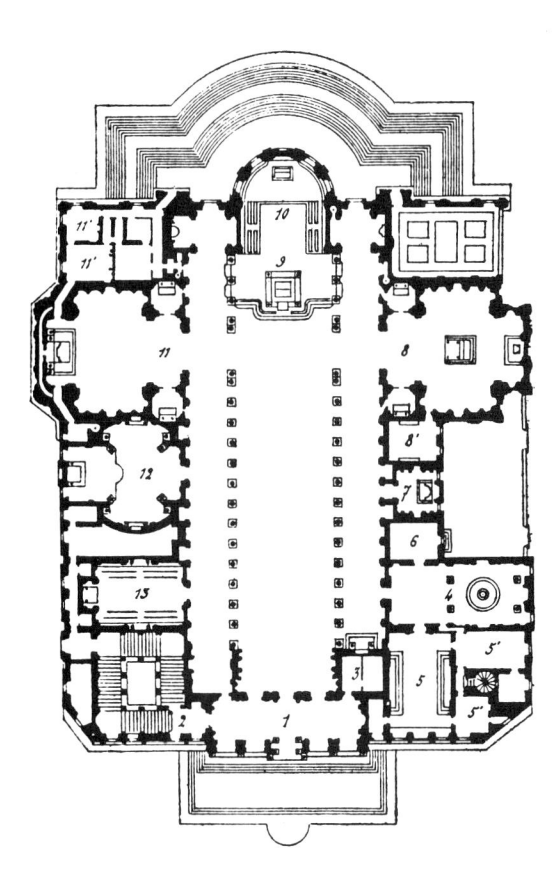

1.42 Plan of Santa Maria Maggiore

steps is an act that they must submit to before entering the house of God (or the palace of temporal rulers). The steps insist that we participate in the ritual of approach.

THE SYMBOLS OF POWER

Secular demonstrations of the monumental stair are every bit as common as those erected for spiritual edifices, for wherever autocratic power is exerted over large building complexes, there flourishes the monumental stair as an immediate exhibition of the puissance of the king, the empire, the state, and latterly the corporation and institution.

Marwick (1888, 6) describes an early example—Xerxes' Persepolis (fig. 1.43) at the height of the Persian empire:

Large external flights of stairs existed . . . they consist of immense single or double flights of steps leading to artificial, elevated, level plateaux of great height and extent, on which their principal buildings were erected. They formed an approach of the most majestic grandeur, while the buildings themselves were greatly enhanced in architectural effect and nobility by their carefully studied situation. The walls of these staircases were in many cases richly decorated—in some instances with such multitude of figures and mythological representations as to absolutely bewilder the onlooker. The immense double staircase leading to the platform whereon stood the Great Hall of Xerxes (521–486 B.C.) was thus decorated with magnificent bassi-relievi of bulls, lions and figures.

This is a very different attitude from that of Athens in the golden age, for the Athenians had little taste for monumental stairs in their deliberately informal building compositions. The approach to the propylaeum of the Acropolis, for example, was by a

The Architecture of the External Stair

1.43 *Xerxes' Persepolis*

Straight Flight Stairs

winding path rather than by the more formal stairs added later by the Romans. The vast 175-foot-wide flight of the propylaea of the great complex of the temple of Jupiter at Baalbek, Lebanon, built in Hadrian's time, might well have seemed ostentatious, vulgar, lacking in subtlety, aesthetically monotonous, and intellectually repellent to the Athenians. However, to the Imperial Romans (just as much as to the Imperial Persians), if it was right and reasonable to exploit the devices of monumentality, superhuman scale, axiality, and symmetry to the greater glory of the gods of Rome, then why not for the palaces, monuments, and fora of the gods' Rome.

THE FORMAL APPROACH

The external approach stair became a favorite architectural device for emphasis and aggrandizement—not only for monumental buildings but equally for smaller buildings with pretensions, wishing to accentuate their entrances.

The approach stair functionally may be no more than a useful way of reaching the piano nobile without the spatial demands of a noble, internal stair. Palladio's rotunda at the Villa Capra, Vicenza (fig. 1.44), for example, achieves this, although four of these external stairs are provided. This axially symmetrical arrangement proclaims that the villa is perfectly complete from any viewpoint and does not favor any facade or direction. The diagonal thrust of the stair accentuates the entrances and emphasizes the separation of the house from the ground plane almost like a drawbridge or a moat—typical of the Renaissance fascination with the relationship between interior and exterior spaces (an interest shared with twentieth-century architects). For the Renaissance architect, the transition from exterior to interior was something to be treated as carefully as any individual space. The act of entry was as important as entering, in much the same way as the act of ascending the stair was as meaningful as reaching the top. Villa Capra, squatting on the ground without its stair approaches, would be as mean as Florence Cathedral, Santa Maria del Fiore, without its dome.

Perhaps one of the most memorable approaches to any group of buildings is Michelangelo's stepped ramp (*cordonata*) for the Capitoline Hill in Rome (1538–1578, approx.)(fig. 1.45). Just as his Laurentian Library stair was the parent of a generation of baroque stairs, the Campidoglio is an early demonstration of baroque planning principles, a tour de force replete with a succession of visual sleights of hand as subtle as any practiced on the Athenians by Ictinus and Callicrates, the designers of the Parthenon.

Michelangelo uses a grand approach stairway to form what was to become a model for the typical baroque vista. Viewed from the top of the flight, the vista is exaggerated in scale by a subtle increase in the perspective formed by narrowing the stair as it recedes into the distance. This repeats the arrangement of the two side wings of the Campidoglio, which are not parallel. They are closer together the farther they are from the Palazzo del Senatore. A lesser architect would have hesitated before risking this trapezoid shape; however, Michelangelo was compelled to accept the condition that the quattrocento Palazzo del Conservatore and the medieval Palazzo del Senatore were to be retained and covered with new facades, and they were not at right angles to each other.

The Architecture of the External Stair

1.44 *Rotunda at Villa Capra, Vicenza*

1.45 *Cordonata, Capitoline Hill, Rome*

Straight Flight Stairs

He chose to complete the court with a matching wing on the other side and to accept the trapezoid ground plan.

In an approach from below, the reverse perspective caused by widening the steps and the piazza (contrary to what one might expect from a study of the drawings) does not seem to foreshorten the building visually. Widening the stair simply makes the approach seem precisely sufficient for the scale of the court. A narrower stair would have seemed mean. And widening the court as it approaches the focus makes the space appear somewhat larger than it really is. All of this is achieved almost imperceptibly; the stair does not appear to narrow, and the piazza's sides at first glance appear to be parallel. The approach and the piazza form one of the greatest pieces of architectural theater in a theatrical age.

The climbing pole, ladder, and companionway continue to serve as simple, steep access ways. The straight flight stair seems to be the oldest proper stair, and all the phases of its evolution remain in use today. These stairs have been used to create architectural effects that are equaled and perhaps surpassed only by some of the more complex and extravagant layouts of baroque Europe.

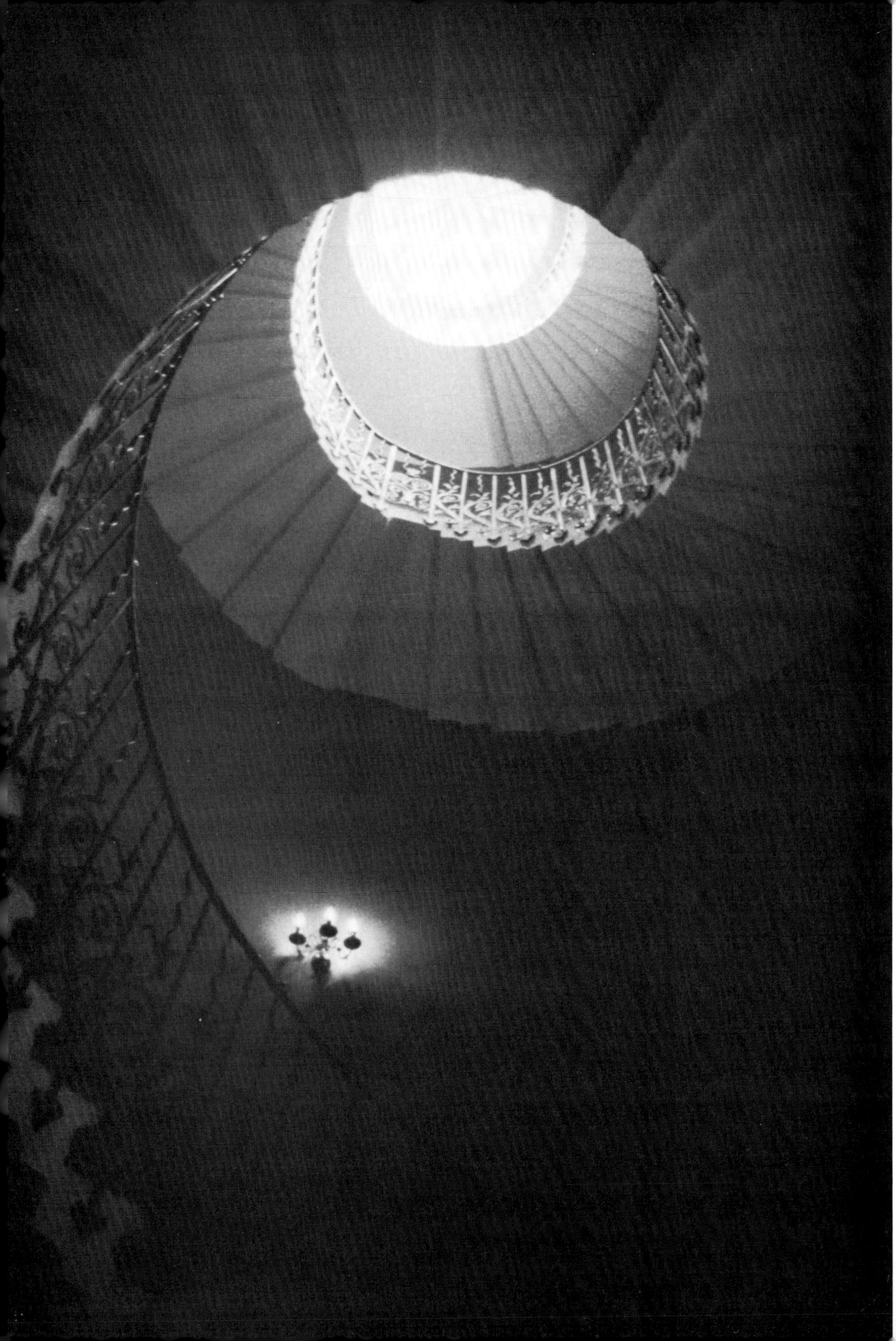

2 THE HELICAL STAIR

This winding, gyring, spiring treadmill of a stair is my ancestral stair.
—W. B. Yeats

All rising to great place is by a winding stair.
—Francis Bacon

The helical stair is also called the spiral stair, winding stair, circular stair, elliptical stair, oval stair, geometric stair, vis, vice, vis de Saint Gilles, St. Gilles screw, belfry stair, turret stair, caracole, turnpike, cochlea, cockle, corkscrew, and ascensorium. *Helical* is possibly the least poetic but also the most accurate generic title.

It is an ancient device and was certainly known in biblical times. From the Old Testament we learn that in Solomon's temple (970 B.C.), "they went up with winding stairs into the middle chamber" (1 Kings 6:8). There were helical stairs in the palace of Diocletian at Spalato (A.D. 300). Even earlier, Trajan's column (A.D. 113) in Rome had a helical stair carved out of solid white marble blocks, 12 feet (3.66 m) in diameter and 5 feet (1.5 m) high. Justinian's San Vitale, Ravenna (A.D. 526–547), uses the helix (fig. 2.2), and so does its architectural progeny, Charlemagne's Palatine Chapel at Aachen (A.D. 805). However, it was in buildings of all kinds constructed after the dissolution of the Roman Empire that the helical stair became common. The famous

2.1 Tulip stair, Queen's House, Greenwich, England

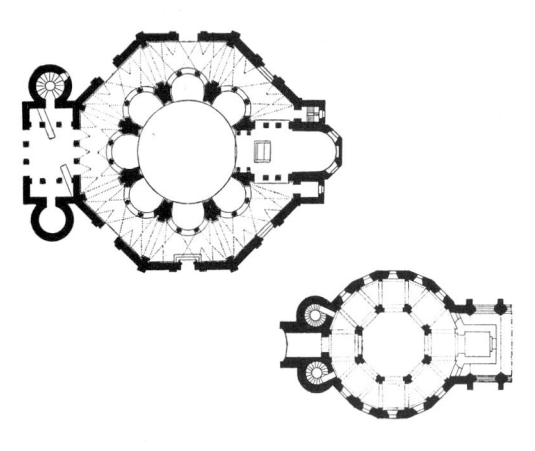

2.2 Plans of San Vitale, Ravenna (top), and the Palatine Chapel, Aachen (bottom)

2.3 Vis de Saint Gilles, near Arles, France

and influential Abbey Church at Cluny (1088–1118) had five or six leading to its towers, the Church of the Apostles, Cologne (1035–1220), has four, as does Worms Cathedral, and Durham (1093–1133) and Notre Dame (1163–1256) cathedrals both have six.

The helical stair is an appealing design because it occupies less floor space than other layouts; however, its design and construction require the technical skills of a craftsman, so it is not surprising that the epoch of the helical stair in the great castles, churches, monasteries, and palaces of Europe coincides with the development of the craft guilds in the Middle Ages. The Vis de Saint Gilles, in fact, completed in 1142 as part of Saint Gilles du Gard near Arles in Provence, gave its name to the genus. The exquisite geometry and quality of the stone steps, vaulting, and newel attracted craftsmen from all over Europe, who marked their visit to the stair with engraved graffiti to pay homage to it (fig. 2.3). Student masons also made the pilgrimage and made models of the stair as their masterpiece for their masters. The bell tower north of the abbey, which the stair served, is now only a ruin, but fifty of the steps and the vaulting are still intact.

2.1 THE DEFENSIVE VIS

Besides providing access, within the medieval fortress the helical stair was useful for defense. An enemy who succeeded in storming the outer walls of a castle still had to take the fortress. The internal communication system was devised to give the defenders every advantage. Routes through these castles are devious and comprehensible only to those familiar with them. Stairs

The Helical Stair

rise one floor and then continue upward in a different location. They are narrow and precipitous and can be blocked with little trouble. A few armed men at the landing could keep at bay a strong force of attackers, compelled to ascend in single file. If the enemy captured the stair, they would be menaced by overlooking shot holes.

There is a persistent theory that these castle staircases are usually made as dextral helicals, descending to the left—anticlockwise—to give the defender the advantage: "Now, a man in ascending these [sinistral stairs] would naturally cling to the column . . . leaving his sword hand free, while the defenders descending the staircase would have their right hand fumbling along the newel" (Rothery n.d., 50). This would not help much if the walls had been escaladed and the defenders were forced to retreat downward or if the defender was left-handed. There are also stories of stairs being built to match the sword hand of their noble owner. This functional argument may be tenable for small stairs but not for the grander medieval stairs. For example, Orford castle, a well-preserved Norman keep (1172) on the Suffolk coast of England (fig. 2.4), has a 12-foot-diameter (3.66 m) stair with treads that are about 63 inches (1.6 m) wide. Seeking the support of the newel on such a wide stair would be decidedly risky; one would have to walk on the narrowest part of the treads, and a loss of balance could precipitate a 6- to 8-foot (2–3 m) vertical fall, so it is unlikely that the defender would choose this position willingly. On the other hand, the typical helical stair tread in these castles is less than 3 feet (1 m) wide, so the tread depth increases greatly as the

2.4 Great spiral, Orford Castle, England

2.5 Dover Castle, England

distance from the newel increases. Standing in the middle of the tread gives plenty of room for one's feet and makes it possible to hook the left arm around the newel. However, the newel is not a column; it is never a complete circle, so it is not easy to hold onto it.

The medieval helical stair offered other advantages to the military architects. In the castle war machine, space was extremely precious. Every square foot of internal space demanded a heavy investment in external perimeter walling and outworks, and the helical stair occupies less space than any other. For efficiency, they were sometimes constructed outside the walls, in turrets. Often they were built within the massive walls of castles, as at Dover (fig. 2.5), where, because of their shape, they weaken the carcass under attack less than orthogonal layouts.

2.2 THE HELICAL STAIR AS AN ARCHITECTURAL DEVICE

Most helical stairs embedded within the massive walls of medieval castles are precipitous, dark, and uninviting. Some are nevertheless quite large and decorative, responding to the puissance of the owner. Hedingham castle, a Norman keep built between 1130 and 1152 in Essex, has a 14-foot (4.25 m) diameter stair leading to the battlements. Much of it is built within the thickness of its 14-foot (4.25 m) thick walls (fig. 2.6).

As military architecture evolved, from the beginning of the thirteenth century onward, impediments to the enemy's approach were no longer constructed inside the keep. The staircase became more an instrument of communication, embellishment, and even spectacle than of defense.

The Helical Stair

2.6 Hedingham Castle, England

For example, in 1364 Raymond de Temple, master of works to Charles V, in his transformation of the old Louvre in Paris from a medieval castle into a palace, erected a famous 27-foot-diameter (8.22 m) staircase (la Grande Viz Neuve, demolished in about 1624).[1] These grander castle stairs were emancipated from the walls and enclosed in towers or turrets constructed solely for this purpose. Because of this, they have more light and air. The stairs are usually circular in plan regardless of the shape of the turret, though examples of square and polygonal circumferences exist (Penshurst Place in Kent has a polygonal outline matching its turret).

The three round towers on the Quai de l'Horloge (named for the first public clock in Paris, made during the Renaissance) typify the turret stairs of the Middle Ages (fig. 2.7). They are all that remains of the palace built by Philip the Fair (1285–1314) on the Île de la Cité. The building was later turned into the Palace of Justice, housing the Conciergerie. From the guardroom of the prison, Marie Antoinette and more than two thousand other condemned prisoners of the Revolution climbed the spiral staircase on their journey to the cells and execution by the Revolutionary Tribunal.

As these turret stairs became part of the compendium of building elements, designers began to experiment with them (in much the same way as exterior fire stairs have evolved aesthetically in the twentieth century). We find a stair spiraling down next to a church tower of the Kilianskirche at Heilbronn (1513–1529) (fig. 2.8) in the same aesthetic relationship that Le Corbusier uses in the Unité d'habitation in Marseilles (fig. 2.9). The twisting, winding shape contrasts with the static mass of the

The Helical Stair as an Architectural Device

2.7 Medieval towers of the palace of Philip the Fair, Paris

The Helical Stair

adjoining building. In both instances, the stair is consciously exposed, not concealed within the building, and the stair's sculptural qualities are exploited.

POWER AND STATUS

In the buildings of medieval cities, helical stairs were popular. They occupied little space, the entry and exit points were not far apart, as they were for straight flight stairs, so finding a place to locate one was relatively simple. Helical stairs were preferred for other reasons as well. They were fashionable and increasingly used as demonstrations of the status of the owner in much the same way as the great baroque stairs of the seventeenth and eighteenth centuries were made the focus of the palace. As status symbols, the turret of the fortress and the bell towers of the church

2.8 Kilianskirche, Heilbronn, Germany

2.9 Unité d'habitation, Marseilles, by Le Corbusier

The Helical Stair as an Architectural Device

became favored architectural devices for palaces, patrician houses, and other similar types of buildings, symbolizing in the city landscape the family's power and position. Some Italian towns such as Pisa, Bologna, and San Gimignano had hundreds of towers built by powerful families.

The stair tower became a familiar feature of civic buildings. In the town hall in Rothenburg (fig. 2.10), for example, the stair tower has the most prominent position in the facade and is the focus of the symmetrical composition. Walter Gropius did the same when he used a pair of stair towers to anchor the two ends of the symmetrical facade for his 1914 Werkbund exhibition building in Cologne (fig. 2.11).

Emancipation of the Vis; Dissolution of the Walls

Internally, also, helical stairs began to receive special attention. They were transformed from dismal holes into sculptural objects, new spatial experiences, and symbols of pride and prestige. These stairs have more light and can be seen as expressive elements. To achieve this, the stair was revealed by removing much of the surrounding walls.

The walls of exterior stairs were often replaced by the flimsiest of structural cages. The Campanile at Pisa Cathedral (c.1174), in essence a 52-foot-diameter (15.84 m) freestanding helical stair, is one of the earliest examples. At Strasbourg Cathedral (1277–1438), four stairs stand free from the tower at four corners and rise up in open, light, airy towers to the level of the base of the spire (fig. 2.12). The stair towers become a supporting skeleton, matching the Gothic motif of general wall dissolution. The cathedrals at Prague (1372) and Rottweil (1330–1479) have open stair towers also, comparable in feeling to Arne Jacobsen's external emergency stair for a factory building at Copenhagen (fig. 2.13). Glass enables the latter stair to have the openness and dematerialization sought by the Gothic architects and to be protected from the weather.

The great stair built by Francis I (1515–1525) at his castle at Blois (fig. 2.14) is perhaps the finest example of the grand helical stair, with dissolved walls, attached to the facade of a building. Its conception and detailing are clearly the work of a master, and there is at least strong circumstantial evidence that Leonardo da Vinci was the designer. Construction of the stair started within a year of Leonardo's employment by Francis I, and he lived at Clos Luce (or Cloux), less than 20 miles from Blois. Previously Leonardo had worked closely with Bramante, whose helical ramp for the Belvedere was so influential (figs. 2.15 and 2.16). And the Blois stair is articulated from the facade in the manner of Bramante's and Leonardo's drawings of the period. Leonardo's notes and sketches show his interest in stairs and shell shapes similar to those applied at Blois. His interest is practical as well as creative. For example, for public stairs leading to the streets, he recommends the use of a spiral staircase and comments, "It should be round because in the corners of square ones, nuisances are apt to be committed."

The stair at Blois is quite revolutionary in its deliberate articulation and because it is an open stair. The view from the stair is outward rather than inward toward a grand stairwell, which was more typical of the work of Bramante, Palladio, Bernini, and Vignola. The stair user experiences a

2.10 Rathaus, Rothenburg, Germany

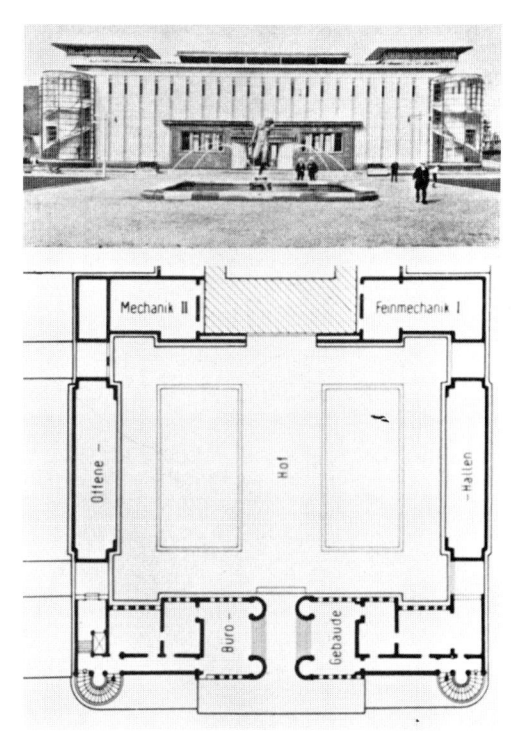

2.11 Werkbund exhibition, Cologne, 1914, by Walter Gropius, reprinted from Mielke (1966), 55

The Helical Stair as an Architectural Device

2.12 Strasbourg Cathedral's four helical stairs

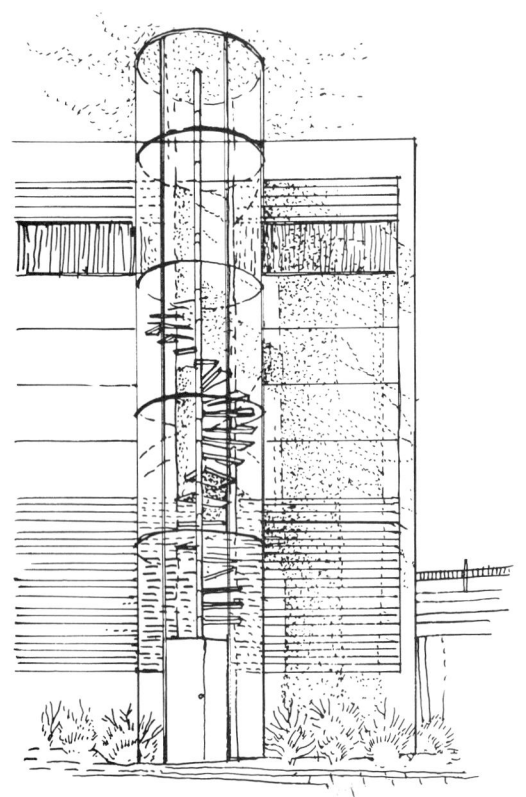

2.14 Blois, the Renaissance stair

2.13 Jacobsen's factory, Copenhagen

The Helical Stair

2.15 Bramante's Belvedere ramp in the Vatican

2.16 Plan and section of the Belvedere ramp

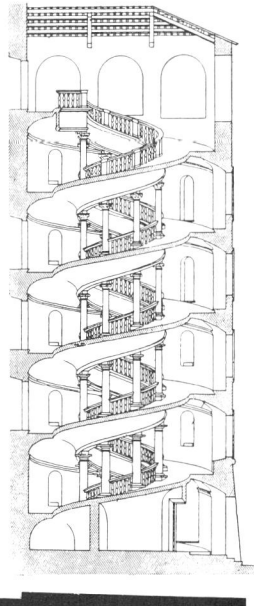

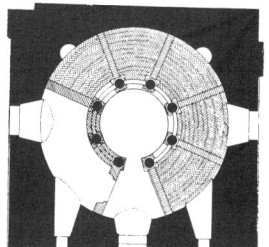

The Helical Stair as an Architectural Device

changing exterior view in much the same way as John Portman was later to introduce in his use of glass elevators. This is a dramatically richer spatial experience than that of even the grandest of the medieval vises.

The decorative treatment of the stair is also notable. As the stair winds upward, the decoration on the newel spirals upward too, and the treads spring from the newel in a sinuous curve like the petals of a flower unfolding.

Most of the early helical stairs revolve around a solid newel. For comfort, and to avoid a tread that reduces to unusable dimensions, the newel must be large as at Pisa. An alternative is to make the treads "dance." At the Kilianskirche, for example, the narrow end of the wedge-shaped tread is widened, and the nosings no longer radiate from a common center. The nosing line remains straight until close to the newel, and then it cuts back abruptly, in a concave curve, to the usual position (fig. 2.17). When the tread is longer, the nosing line can be made to dance by bowing out more or less abruptly from the newel and then curving back again toward the outer string. At Blois, for reasons that can only be decorative (fig. 2.18), instead of forming an asymmetrical bow, the nosing line is a sinuous S, which widens out again against the newel.

Internally also, from early in the twelfth century, helical stairs began to lose their enclosing walls. Notre Dame (1163) in Paris has two very small open cage stairs—only 4½ feet (1.37 m) in diameter—in its two towers. A similar stair, built in the fourteenth century and restored in the nineteenth, leads from the Hall of the Gentlemen-at-Arms in the Conciergerie in Paris (fig. 2.19). All of these stairs are liberated from the surrounding wall by a simple structural solution: transmitting the

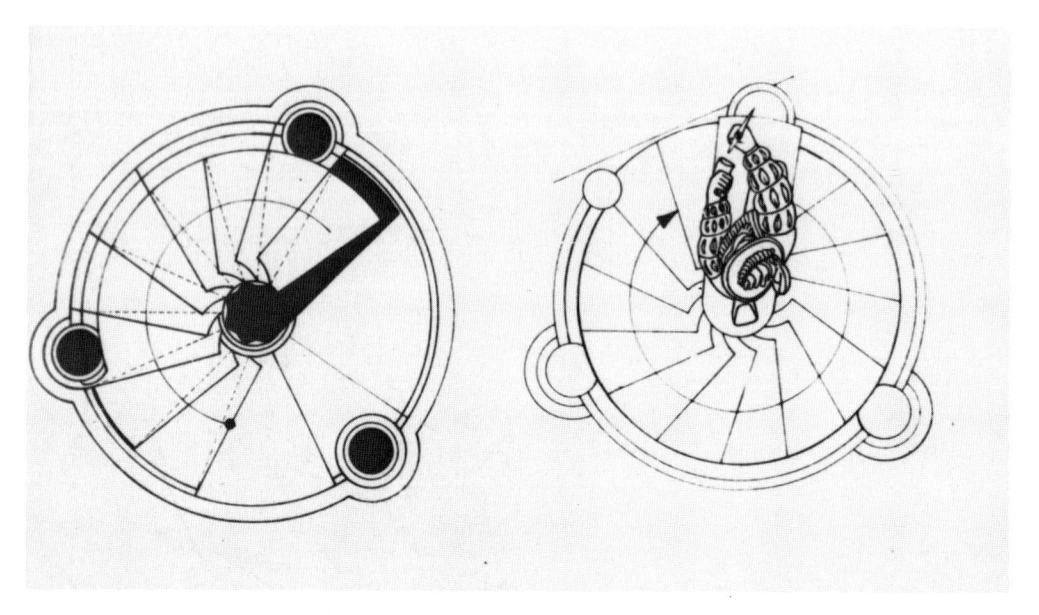

2.17 Dancing steps at the Kilianskirche, Heilbronn, reprinted from Mielke (1966), 50

2.18 Sinuous nosing of the Blois stair

2.19 Fourteenth-century open-cage helical stair, Conciergerie, Paris

The Helical Stair as an Architectural Device

2.20 *Georgenkirche, Nordlingen*

loads vertically through a columnar framework. This allows the stairs to stand free, visually and structurally.

The pulpit became an increasingly important functional element in the church, and the helical stair provided a natural accessway. Pulpit and stair became an expressive sculptural element, providing an incentive for exploring new solutions to the problem of the enclosing wall. Arch and cantilever construction enables the steps to project out from a single support, with no external wall, as in the Georgenkirche in Nordlingen (1499) (fig. 2.20). This type became popular with the development of cast-iron and prefabrication techniques. In the nineteenth century, particularly, there are countless examples of freestanding cast-iron helical stairs revolving around a single column, notably in John Nash's Royal Pavilion, Brighton (1815).

Ultimately even the newel and risers vanish, and the loads are transmitted toward the ground on a spiraling inner string, often with treads cantilevered from it. This is, of course, a favored device of twentieth-century architects, and Niemeyer's Palace of the Arches at Brasília is an elegant example (fig. 2.21).

DISSOLUTION OF THE NEWEL
Dissolving the walls was simply a first stage in the deliberate attempt to explore the spatial possibilities of the helical stair. There followed experiments with the newel, to open up the eye of the stair—and ultimately to strive for a magical stair devoid of walls or apparent support. By dissolving the solid newel into a hollow shaft or frame, one can introduce light from the top down the middle of the stair, a practical discovery for internal stairs particularly.

The Helical Stair

2.21 *Palace of the Arches, Brasília, by Oscar Niemeyer*

The Helical Stair as an Architectural Device

At first these open wells were not much bigger than the old solid newels. In the Frauenhaus (1578–1582) in Strasbourg, the solid newel becomes three small, delicate columns, which continue up beyond the stair to the vaulted roof above (fig. 2.22). A sinuous handrail carved with grooves climbs around the columns like a creeper, paralleling the path of the stair, and emphasizes the spiraling movement. In a more daring version, the single newel forms a spiral like a twisted elastic, and then the whole newel appears to wrap itself around an invisible core as it ascends. At the Gemological Institute building, Los Angeles, of Richard Neutra (fig. 2.23), this device creates a dynamic tension from its apparent instability, a game that has attracted many architects. These stairs emphasize the idea of endless movement implicit in their spiral form and reflect the complex spatial path that the user experiences.

Opening the eye of the stair to form a well offers several other benefits. Besides introducing light, the well allows the helical stair to become less constricted and more comfortable to use. The inner portion of the tread increases as the well expands, until eventually it is large enough to walk on comfortably.

Spatially, the well was a revolutionary discovery, connecting the horizontal floors in quite a new way and making stair use a dramatic architectural experience with its increased sense of verticality felt as a continuous spiraling journey. Stairs became as grand as their owners wished and could afford, and the perfecting of the grand helical stair is one of the major architectural achievements of the Renaissance.

Vasari (1511–1574) (1946, 12) attributes the invention of the stairwell to Niccolò of Pisa in his design for the campanile of San Niccola in Pisa: "Externally this building [campanile] has eight sides, but within it is circular, with a spiral staircase. Within the stairs a free space is left like a well, while on every fourth stair columns support arches which follow the spiral line. The roof of the staircase is supported on these arches. The ascent is of such sort that the spectator at the foot sees all who go up, and those above see those remaining below." The stair of San Niccola still exists (fig. 2.24), quite close to its distinguished leaning cousin (and it also has a noticeable lean). It is actually no more than a service stair, and it is gloomy and provides a home for many pigeons. It is unlikely that this unpretentious and largely unknown example served as a model for future generations. Furthermore, Vasari was incorrect; there are earlier examples. The stairways in the Hellenistic towers at Aghios Petros (Andros) and Chimarrou (Naxos) on the Cyclades have open wells, and the steps are formed from blocks of stone cantilevering from the walls (Coulton 1977, 149–150).

It was quite late in the Renaissance, at the beginning of the sixteenth century, that the stairwell for helical stairs was fully developed—much earlier than the stairwell for straight flight or composite stairs. The most influential helical stair (in fact, a ramp) with a substantial well was built by Bramante for Pope Julius II in the tower next to the Belvedere (figs. 2.15 and 2.16). This was soon followed by Vignola's magnificent helical stair at the Farnese Palace at Caprarola (1547–1549), where the stairwell is opened up to almost 10 feet (3 m) (figs. 2.25 and 2.26).

Opening the well presents problems as well as benefits; the greater the well is, the

The Helical Stair

2.22 Frauenhaus, Strasbourg

The Helical Stair as an Architectural Device

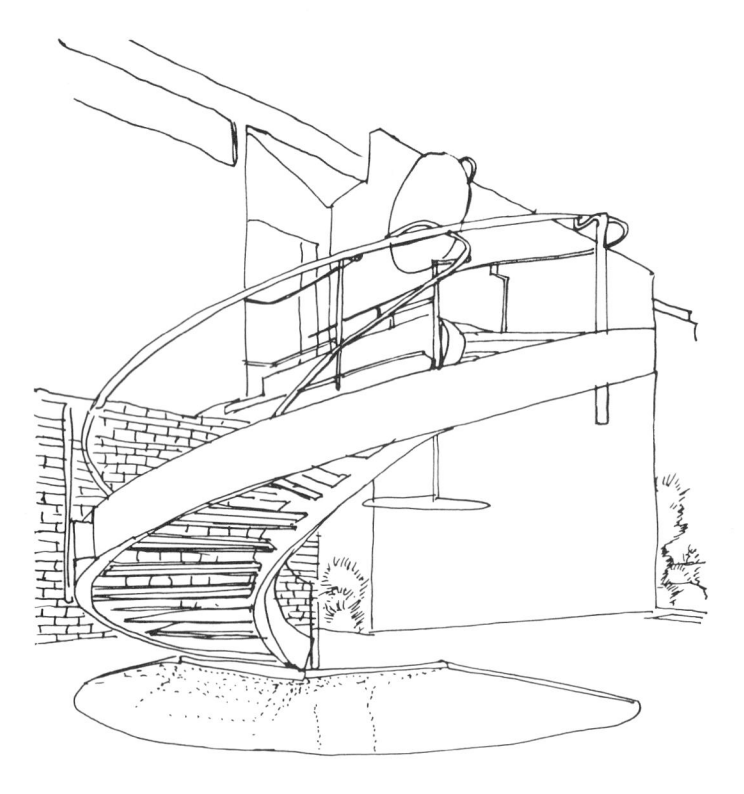

2.23 Gemological Institute, Los Angeles

2.24 San Niccola, Pisa

The Helical Stair

2.25 *Vignola's Farnese Palace, Caprarola*

2.26 *Plan of the Farnese Palace*

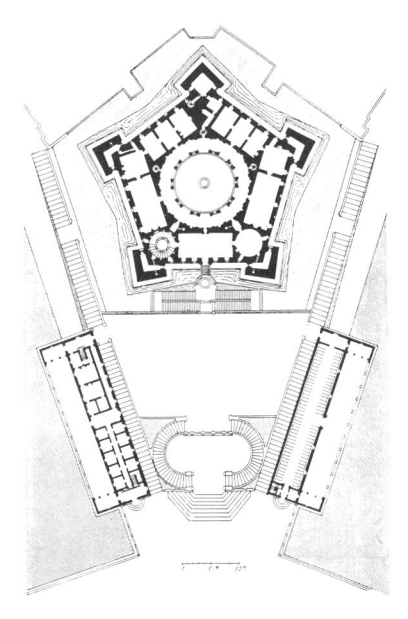

The Helical Stair as an Architectural Device

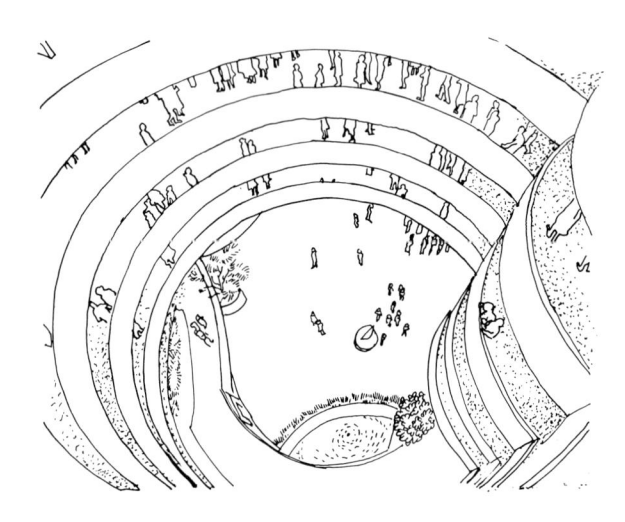

2.27 Guggenheim Museum, New York, by Frank Lloyd Wright

greater is the unused space. And the concept of the great stair hall of the baroque was still in the future. The helical stair and well is like a vertical tunnel of space, with limited possibilities for spatial flow sideways. A spiraling stair forms an apparently continuous wall and strong enclosure. Even Frank Lloyd Wright's Guggenheim Museum in New York City, with its nearly 60-foot-wide (18 m) well and skylit atrium (fig. 2.27), is constrained by these visual, spatial limitations. For these reasons, the baroque designers abandoned the continuous helical stair in favor of designs that were less demanding and limiting spatially.

PERMUTATIONS OF THE HELIX

Palladio appreciated the open well: "I have made a staircase void in the middle, in the Monastery della Carità in Venice, which succeeds admirably" (fig. 2.28). He treats helical stairs in some detail with many illustrations (1570), a marked change from the short shrift meted out to stairs by Alberti a hundred years earlier. During the sixteenth century, stairs in Italy began to develop their central architectural spatial role, which reached its apogee in the baroque and rococo of the seventeenth and eighteenth centuries. Alberti (1986, 19) dismisses stairs as a nuisance: "The fewer staircases that are in a house, and the less room they take up, the more convenient they are esteem'd," he writes.[2] He does, however, suggest that they should be "ample and spatious according to the dignity of the place."

Palladio (1965, 35), on the other hand, devotes a whole chapter, entitled "Of Stairs, and the Various Kinds of Them; and of the Number and Size of the Steps." He applauds the use of "winding staircases,"

The Helical Stair

2.28 Monastery della Carità, Venice

"in narrow places particularly, because they occupy less room than the straight, but are somewhat more difficult to ascend. They succeed very well that are void in the middle, because they can have the light from above, and those that are at the top of the stairs, see all those that come up, or begin to ascend, and are likewise seen by them." He describes with enthusiasm the oval helical, which was just becoming popular: "They are very beautiful and agreeable to see, because all the windows and doors come to the head of the oval, and in the middle, and are sufficiently commodious." He illustrates several of these (figs. 2.29 and 2.30), including that of the School and Church of Santa Maria della Carità in Venice (the Galleria dell'Accademia is now housed in this building), about which he says, "In the part near the church there is an oval staircase, void in the middle, which is very convenient and pleasant."

Borromini's oval helical stair in the Barberini Palace, Rome (1628) (fig. 2.31), closely matches this description, except that the treads have a considerable wash (slope from back to front), like many baroque stairs. At the inside of the stair, because the tread is small, the treads slope rather steeply. Furthermore, walking rhythm on this stair is extremely curious, for the treads change in size as one circulates around the oval form.

A special variant of the circular helical stair was particularly popular in France and Germany, the double helix. Two stairs of equal diameter spiral upward about a common center but start 180 degrees apart. The origins of the double helix are obscure. Mielke (1966, 42) says it is an old Islamic device, noting that there is such a stair in the minaret of the Suq al-Ghash of

2.29 Palladian helicals

The Helical Stair

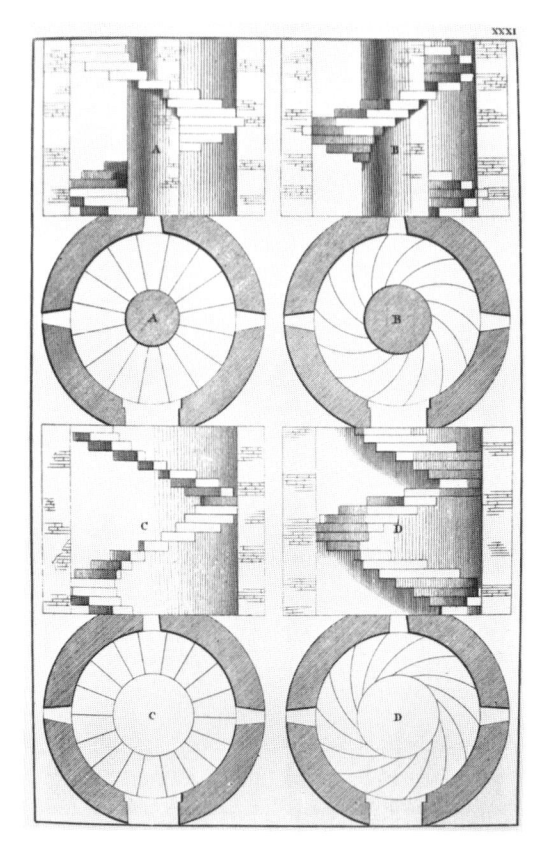

2.30 Palladian helicals

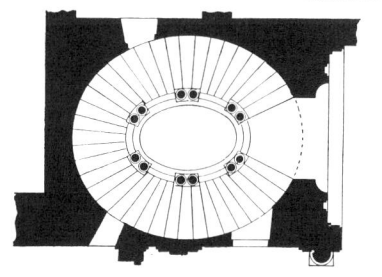

2.31 Barberini Palace, Borromini

The Helical Stair as an Architectural Device

Baghdad (902–908). And Sanval, in his *Histoire et Antiquités de Paris,* speaks of one that formerly existed at the establishment of the Bernardins at Paris. Antonio da Sangallo (1470–1546) constructed an extraordinarily clever double helical ramp for Pope Clement VII at Orvieto in 1527 for animals to use when carrying water from a deep well intended to give Orvieto a water supply in case of siege (fig. 2.32). The design ensures that animals descending to the well for water will not meet those ascending and will not have to turn around at the bottom.

By far the most famous of these double helix stairs is at the palace of Chambord near Blois (1519–1547) (figs. 2.33 and 2.34). It is the hub around which this Renaissance ideal castle revolves. Palladio (1966, 35) knew of this palace and its stair and writes of it thus:

Another beautiful sort of winding stair was made at Chambord (a place in France) by order of the magnanimous King Francis, in a place by him erected in a wood, and in this manner: there are four staircases which have four entrances, that is one each, and ascend the one over the other in such a manner that being made in the middle of the fabrick, they can serve to four apartments without that the inhabitants of the one go down the staircase of the other, and being open in the middle, all see one another going up and down without giving one another the least inconvenience: and because it is a new and beautiful invention, I have inserted it, and marked the staircases with letters in the plan and elevation that one may see where they begin, and how they go up.

Palladio's facts and illustration (fig. 2.35) are not quite accurate. The stair has only two helixes, not four, and certainly it does not serve separate apartments but a cruciform corridor that leads to several apartments (fig. 2.36). This is a plan form that provides the inspiration for Michael Graves's Snyderman house (fig. 2.37).

There is a persistent theory that Leonardo da Vinci designed the Chambord stair. It is certainly possible that he influenced the design, for Francis I's first architect, Domenico da Cortone, was Italian. And there is an extant sketch of such a stair by Leonardo (fig. 2.38) and also a design of his for a group of four independent straight flight stairs climbing around a fixed core (fig. 2.39). Furthermore, Leonardo was patronized by Francis I and lived in the district from 1516, dying there in 1519, the year that work on Chambord began.

Perhaps the most curious variant of the double helix are stairs that revolve literally. These are of wood and, "constructed on a pivot, turn easily, so that by a half revolution all the rooms of a house can be closed by them and passage(s) shut off to apartments to which they before gave access" (*American Architect and Building News*, 1891, no. 811, 21).

Two other helical staircase types are worth mentioning for their ingenuity. The first is a special form of the double helix, where the stair has two or more centers of revolution about which it weaves. There is such a stair at the castle at Graz (1499–1500) (fig. 2.40). The great Neapolitan baroque designer Sanfelice built many complex arrangements using helical components, and he designed a version of the Graz stair. The second type is older; here a minor or service stair spirals upward within the major stair. Chambord follows this pattern; a second stair, 10 feet (3 m) in

The Helical Stair

2.32 Sangallo's double helical ramp, Orvieto

2.33 Double helical stair, palace of Chambord, France

The Helical Stair as an Architectural Device

2.34 Sections through the Chambord stair

2.35 Palladio's idea of the palace of Chambord

The Helical Stair

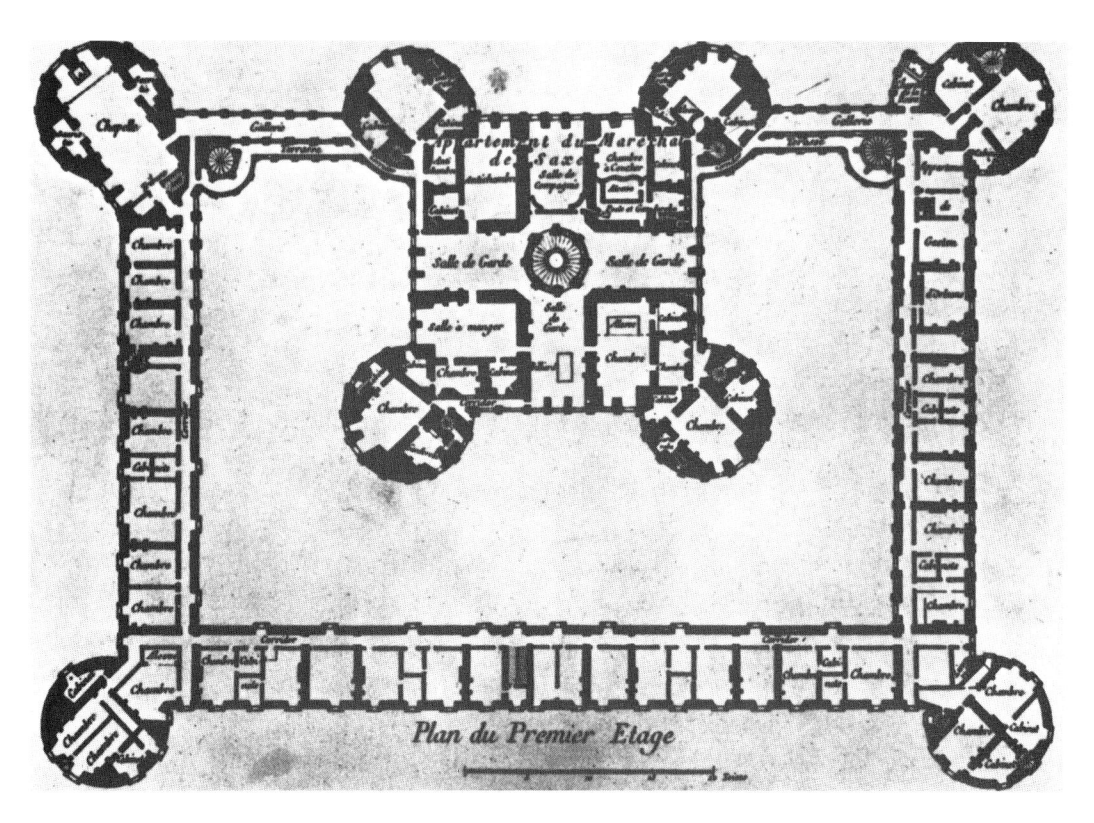

2.36 *Plan of Chambord*

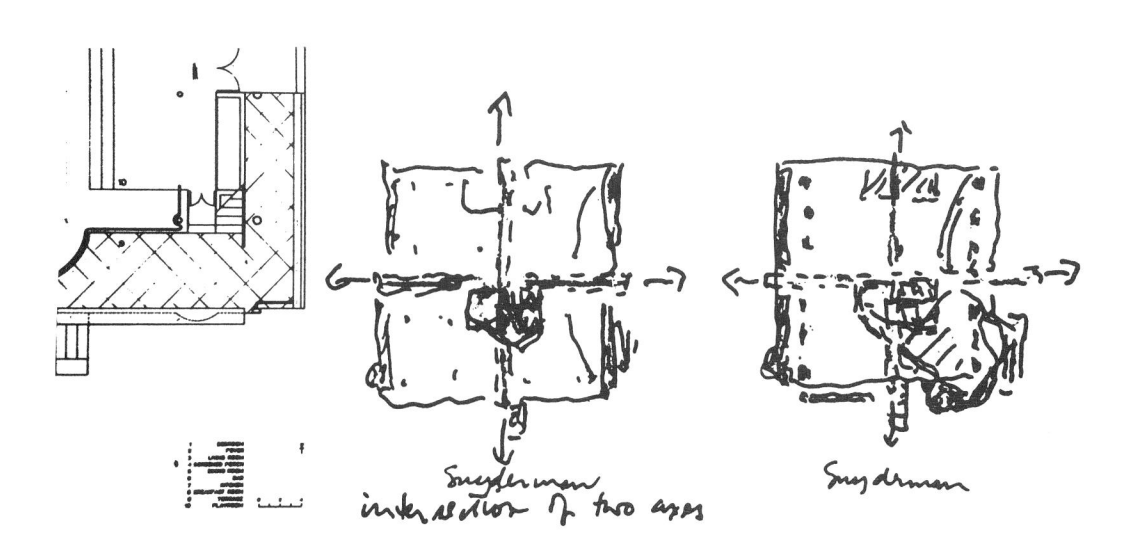

2.37 *Graves's sketch for the Snyderman house, Fort Wayne, Indiana*

The Helical Stair as an Architectural Device

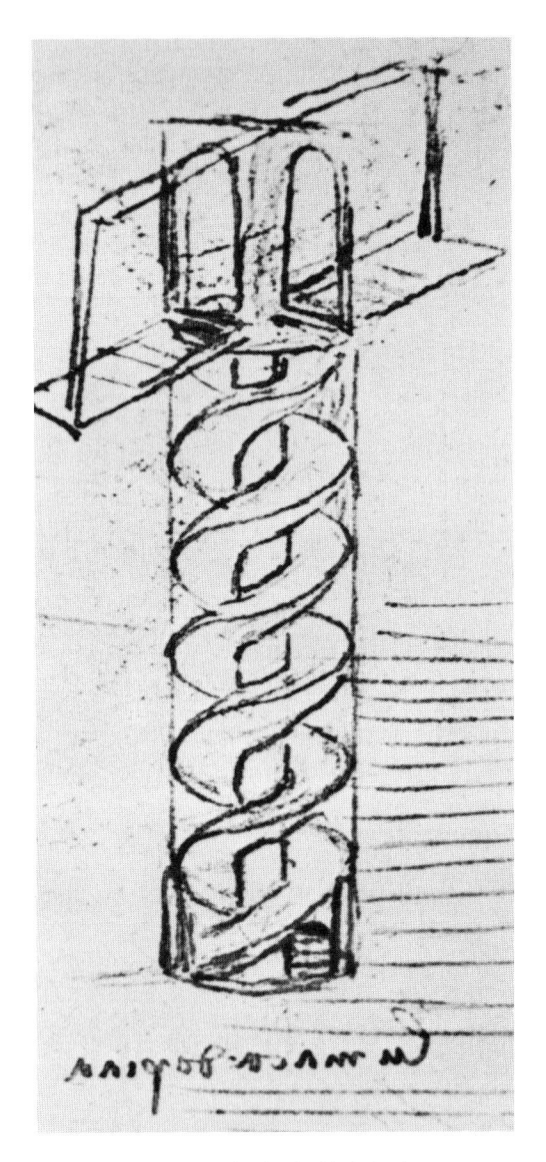

2.38 Leonardo's sketch of a double helical stair

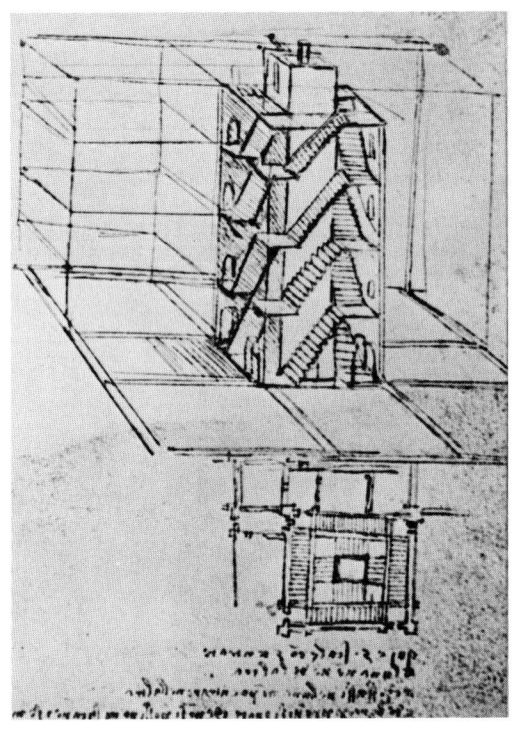

2.39 Leonardo's sketch for four stairs

The Helical Stair

2.40 Double helical at Graz, Austria

diameter, rises from the terrace level up to the tower of the Fleur-de-lys.

Not all helical stairs complete even a whole revolution; the great interest in the stair as a spatial manipulator that arose during the latter part of the Renaissance and the baroque engendered many examples of the partial helix.

The helical stair, like the straight flight, maintains its past in the present. Helical stairs comparable to the medieval vis are still built into the carcasses of ships and factories or wherever else space is scarce; they have solid newels and small radii, and ascent is not particularly comfortable. However, the spatial discoveries of the Middle Ages, when the helical stair lost its enclosing walls and solid newel and took on a gentler rise and larger radius, have gained a new elegance and vigor today, in part helped by the structural abilities of cast iron, steel, and concrete (figs. 2.41 and 2.42). A renewed interest in the architectonic possibilities of the grand stair in compositions has led James Sterling, for example, to use a great helical stair set in a circular unroofed walled courtyard (fig. 2.43) as the focal event of his Stuttgart art gallery. In its relationship to the plan form, it is loosely (and consciously) modeled on K. F. Schinkel's Altes Museum in Berlin (1824–1828).

Much of the baroque fascination for, and evolution of, the grand staircase can be traced directly to the spatial revelations of the helical stair. The helical stair was to medieval and Renaissance buildings what the grand stair was to the baroque, and it was a major element of some of the grandest grand stairs.

The Helical Stair as an Architectural Device

2.41 *Stair at Villa Savoye at Poissy, by Le Corbusier*

2.42 *I. M. Pei's helical for the Louvre museum*

The Helical Stair

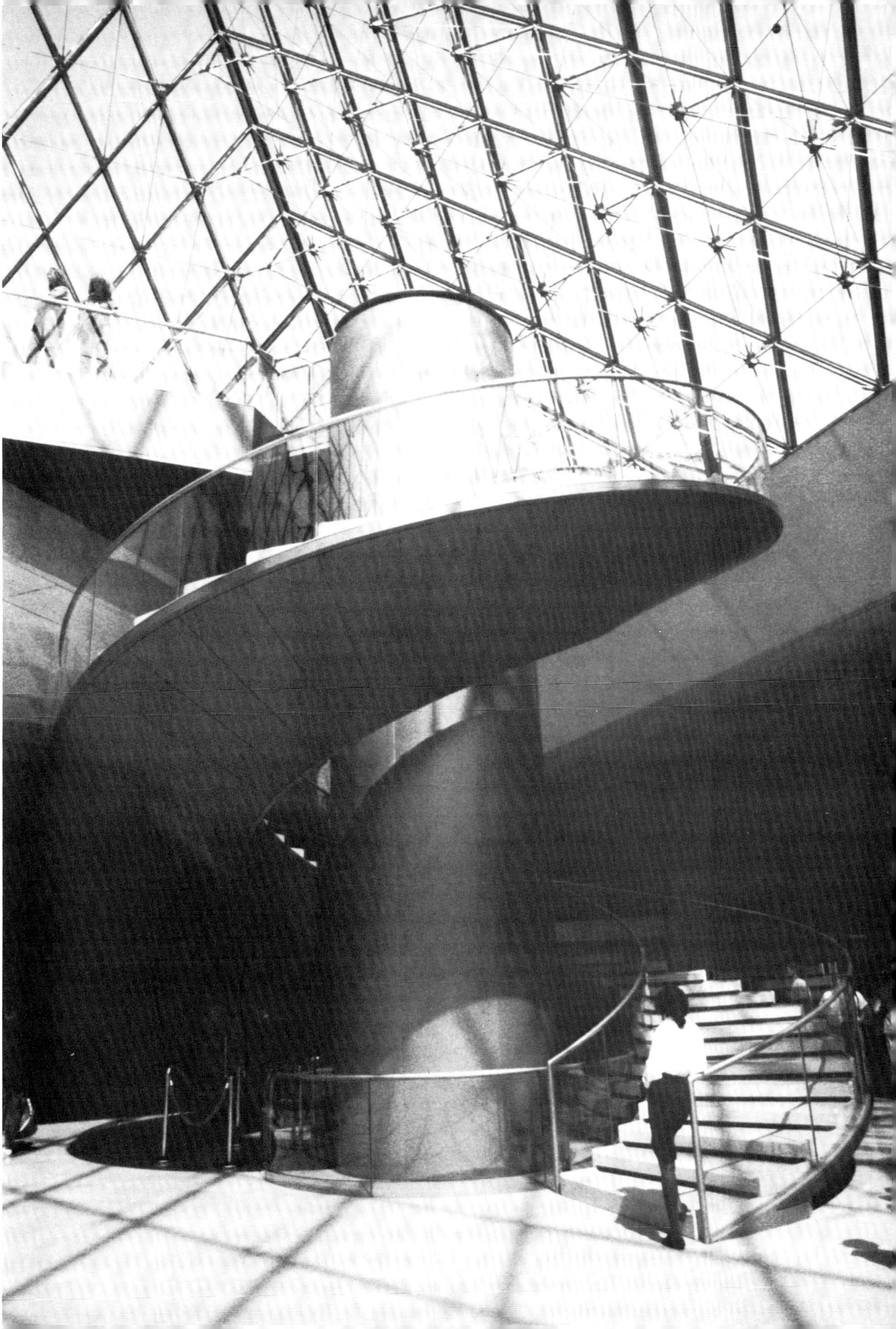

2.43 *Stuttgart art gallery, by James Stirling*

3.1 *Schloss Solitude, near Stuttgart, Germany*

3 Composite Stairs

Helical or straight flights with landings as couplings are the elements from which composite stairs are formed. A dogleg stair consists of two straight flights and a landing; an imperial stair has two parallel straight flights separated from a third by a landing. Even the serpentine flights of Schloss Solitude near Stuttgart (fig. 3.1) are geometrically composed, from helical segments in this case. The possible permutations using these three primary elements are almost infinite. There are some combinations that have been used frequently enough to be considered typical, and there are others that are extraordinarily beautiful or significant in the stream of discovery.

Chronologically, we can trace architectonic mutations from elemental straight flight or helical stairs, through dogleg, open well, and square newel stairs, through garden and grand approach stairs, to the great imperial stairs of Europe.

3.1 Landing Places

Staircases formed of a single very long flight are comparatively rare. Without landings, the daunting flight of the 124 steps leading to Rome's Santa Maria in Aracoeli (fig. 3.2) would be inconceivable perceptually and almost impassable physically for any but the fittest. In very long flights, where the steps are too numerous to be counted in a single glance, the land-

3.2 Santa Maria in Aracoeli, Rome

ings help to provide the visual clues to distance and size. The landing place (to use the whole English phrase) is called in French a *palier de repos* (level of repose). Alberti (1955, 19) proposed a frequency for landings based on symbolic, physiological, and humane considerations:

And I have observed, that the best architects never put above seven or at the most nine steps together in one flight; imitating I suppose the number either of the planets or of the heavens; but at the end of these seven or nine steps, they very considerately made a plain, that such as were weak or tired with the fatigue of the ascent, might have leisure to rest themselves, and that if they should chance to stumble, there might be a place to break their fall, and give them means to recover themselves. And I am thoroughly of opinion that the steps ought to be frequently interrupted by these landing places.

Palladio (1965, 34) picks up the theme, absent any symbolic meaning, but suggests less frequent landings, perhaps reflecting the increasing interest in stairs demonstrated by the gentler gradients and lower risers that began to appear for prestigious stairs: "The number of steps is not to exceed eleven, or thirteen at most, before you make a floor or resting place, that the weak and weary may find where to rest themselves, if obliged to go up higher and be able more easily to stop anything that should happen to fall from above."

These landings not only divide straight flight stairs into areas of activity and pause; they may also provide access points. Alvar Aalto at the Baker House dormitory at MIT in Cambridge (1947–1948) uses, and expresses externally, a continuous straight flight stair from the top to the bottom of the building. It is divided only by landings,

Composite Stairs

giving access to the floors that the stair passes (fig. 3.3). This design is emulated in the escalator access at the Pompidou Center in Paris (fig. 3.104).

3.2 THE DOGLEG STAIR

Probably the composite stair type with the longest continuous history is the dogleg. The dogleg (or dog-legged) stair has two flights, parallel to each other and connected by an intermediate landing, with a half-turn in the direction of travel at the landing. From this landing, the users may ascend or descend and exit in the direction from which they entered. There is a wooden staircase of this type leading from the peristyle court to the upper chambers of the House on the Hill at Delos (second century B.C.). Many of the buildings of the Roman Empire also used these double returning flights, separated by a central wall.

In post-Roman Europe, the dogleg stair was seldom used. The straight flight and the helical were the stairs of choice until the advent of the Renaissance. An exception is the Escalier d'Honneur built by Clement VI for the New Palace of the Popes at Avignon (1342–1352). This stair capped with ribbed vaulting leads down to the Grand Audience Chamber (fig. 3.4).

3.3 Baker House, MIT, Cambridge

3.4 Escalier d'Honneur, Palace of the Popes, Avignon

One of the earliest uses of the dogleg in a major Renaissance building was in Brunelleschi's Foundling Hospital in Florence (1421–1445).

By the middle of the sixteenth century in Italy, the staircase was recognized, finally, as a major architectural device. In the 1568 edition of his *Lives of the Artists,* Vasari for the first time interests himself in stairs and describes them as "the arms and legs of the building's body." And Vincenzo Scamozzi (1552–1616) writes, "Beyond all doubt it is the stair, of all architectural elements, which is for buildings what the blood vessels, arteries, and veins are in the human body: just as the latter bring blood to all the limbs, so the former, designed with similar artistry and similarly ramified are necessary for communication."

During the Renaissance, the dogleg replaced the helical as the typical stair. The Scala d'Oro (figs. 3.5 and 3.6) of the Doge's Palace in Venice (1550) and that of the Farnese Palace by Sangallo in Rome (1530) demonstrate the interest in this relatively new stair type.

The dogleg stair was interesting to Renaissance designers as an ordering device for the plan. It enables the exit and entry points to be located almost directly above each other, permitting floor plans to be regularized with rooms over rooms and the whole plan to be repeated if necessary. This was particularly useful for the ideal system of repetitive structures and spaces characteristic of Renaissance and baroque planning.

The dividing and enclosing walls so typical of these early Italian Renaissance stairs fitted well with the extant conception of harmony derived from the theory that architectural spaces should be complete in

Composite Stairs

3.5 *Scala d'Oro, Doge's Palace, Venice*

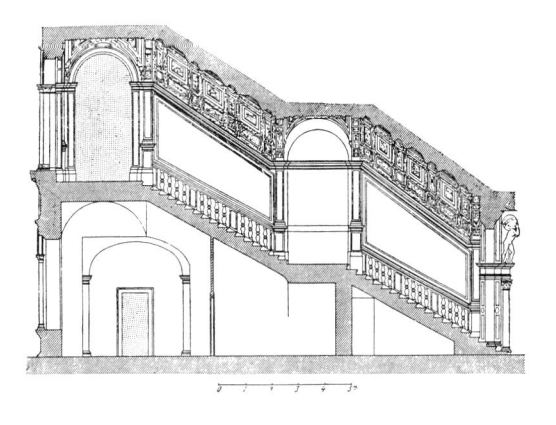

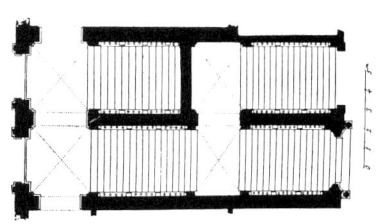

3.6 *Plan and section of the Scala d'Oro*

The Dogleg Stair

themselves. The stairs could be used to give a controlled perspective view, as well as rich, continuous experience. They could be considered as part of the sequence of the processional way and the reception spaces.

This Renaissance theory of space is perhaps the reason that dogleg stairs did not lose the dividing wall between adjoining flights until the sixteenth century, while helical stairs lost theirs a century earlier. It was the dissolution of these walls that finally changed the character, if not the nature, of the dogleg stair. It released its possibilities as a visually continuous route, as a spatial connector, and led to the great open well stairs of the baroque.

Even the service dogleg stairs of today differ little from their predecessors conceptually. Leonardo da Vinci, among his numerous sketches of stairs, even illustrates the scissor stair (fig. 3.7), which has become a favorite device of twentieth-century fire marshals. Stylistic differences are observable between the dogleg stair of various periods of architectural fashion. Victor Horta's rue Paul-Emile Janson house in Brussels (1892–1893), for example, superimposes the characteristic art nouveau decorative features onto a basic dogleg stair (fig. 3.8). The same is the case with Rudolph Steiner's Goetheanum at Dornach (1925–1928), where the stair and hall (fig. 3.9) appear to be carved out of the monolithic concrete, and light from stained glass windows casts an almost medieval air of mystery and spirituality onto what is in essence still a traditional stair.

3.3 FROM SQUARE NEWEL TO IMPERIAL STAIR

By the sixteenth century, the spatial economy and architectonic simplicity of the dogleg stair was no longer adequate for the

3.7 Sketch of scissor stairs by Leonardo da Vinci

3.8 Victor Horta's Janson house, Brussels

Composite Stairs

3.9 *Goetheanum, Dornach, Switzerland*

noble buildings of Europe. Changing theories of the ceremonial use of space led to a desire to treat the stair with as much architectural respect as any other significant area in the building. The rectangular (including square) newel stair became the first product of this new direction. (The square newel staircase has three or four straight flights, coupled by corner landings and enclosed by a wall. The flights climb around an open well.) There were stairs of this general design in Roman times; however, the central well in these cases was solid and not included in the perceptual spatial volume of the stair.

At the Palace of King Minos at Knossos, built about 4,000 years ago, a grand dogleg staircase leads to the state rooms (fig. 3.10). A light well adjoins the stair, giving it a feeling of spaciousness and a grandeur quite unobtainable in the traditional enclosed dogleg. The stair surrounding a light well developed much later and did not appear with any frequency until early in the sixteenth century in Europe. The rectangular open well stair passed through the same transitional process in the development of the open newel as the helical stair. Francesco di Giorgio's treatise of about 1480 (Martini 1967) was the first to illustrate the square open newel staircase.

The open stairwell was rarely used for French or Italian Renaissance staircases. It was in Spain that most of the significant architectonic developments of the stairwell emerged (Wilkinson 1975, 65–90). Perhaps the first of these rectangular open newel stairs was that of the Monastery of San Juan de Los Reyes in Toledo (fig. 3.11), designed by Juan Guas and completed by Enrique Egas in about 1504. This stair had many near-contemporaries, including the Palace of La Calahorra by Lorenzo Vásquez and others (1511), the Mendoza Hospital of Santa Cruz (c. 1530) in Toledo (fig. 3.12) by Enrique Egas with the stair by Alonso de Covarrubias, and the Colegio de los Irlandeses in Salamanca (1535) by Diego de Siloe and Covarrubias (fig. 3.13). All three of these stairs (and many of their successors) open directly off a cloister of a courtyard. Each has three flights and follows three sides of a rectangle, leaving a double-volume open well in the middle. The fourth side opens onto the cloisters at the top and bottom. On three sides, the stair is enclosed by walls. (The Salamanca stair has a balustraded arcade at the upper level on these three sides.) The open stairwell ensures that the whole stair can be seen at a glance, giving it an unprecedented significance and appearance in the building. The stair visually and functionally connects the two levels with a spatial generosity that signals its importance within the building composition. Each of the stairs is treated as a decorative element and is adorned with elegant columns and balustrades.

The rectangular newel staircase also became popular in Tudor and Jacobean England, usually framed in wood and with rather modest dimensions. Hatfield House in Hertfordshire (1608–1612), designed by Robert Lyminge (with perhaps some assistance from the young Inigo Jones), has one of the finest examples (fig. 3.14). It is elaborately carved in oak, and each newel post is surmounted by a figure. Rothery (n.d.) quotes Francis Bacon in his essay on building: "The stairs . . . to the upper rooms, let them be upon a fair open newel, and finely railed in, with images of wood cast in brass colour, and very fair landing place at the top." Godington (1627) and Knole

3.10 Stair at Knossos, Crete

3.11 Monastery of San Juan de los Reyes, Toledo

From Square Newel to Imperial Stair

3.12 Mendoza Hospital of Santa Cruz, Toledo

3.13 Colegio de los Irlandeses, Salamanca

Composite Stairs

(1605–1608) both have examples of these beautiful English wooden stairs, decorated with massive and elaborately carved handrails and newel posts.

The development of the open well stair proceeded with much ingenuity in Spain, and it was largely in Spain and Genoa that the next advances occurred. The rectangular newel stair was a new architectural element, but it was unsatisfactory for the coming age where axially symmetrical compositions were to become a significant part of architectural theory. And, significantly, none of these rectangular newel Spanish stairs was located on an important axis.

The rectangular newel and the dogleg stair posed an architectonic problem. If the entry flight was set symmetrically on the intended axis, the exit flight was off the axis. Alternatively, if the axis was centered on the whole stair rather than on one flight, the stair formed an asymmetrical focus. Neither of these solutions was adequate, and the search began for a symmetrical staircase layout.

Alonso de Covarrubias, the designer of the rectangular newel stairs in the Mendoza Hospital of Santa Cruz in Toledo and in the Colegio de los Irlandeses in Salamanca, appears to have been the first to find a solution. In the Hospital of San Juan Bautista (1541) in Toledo, he gave the stair an importance in relationship to the front entrance that adumbrates the great German baroque stairs: the stair was located centrally on the axis connecting the main entrance to the apse of the hospital chapel (fig. 3.15).

One could enter the stair from flights facing to the left or right of the vestibule. These ended at a landing in the center on

3.14 Hatfield house, Hertfordshire, England

the axis. From there a single flight on the axis led away from the vestibule, ending at a second landing. From that point, flights rose, parallel to the first two flights, emerging into corridors of two adjoining courtyards. Covarrubias used essentially the same overturned H plan for the staircase in the Alcazar in Madrid, completed in 1548 (and destroyed by fire in the eighteenth century).

The elegant symmetry of the overturned H plan staircase is an ideal solution for a bilateral layout, but is weakened conceptually by the initial directional ambiguity offered by the two alternative flights starting on opposite sides of the vestibule. The H plan's success was limited to plans where split termini were desired or acceptable.

The next point on the journey toward the imperial staircase was the development of a symmetrical staircase accessed by a single flight and terminating, after landings, in two flights facing the original entrance—like a pair of rectangular open well stairs set side by side. Two almost contemporaneous examples were constructed in Spain and in Genoa. The staircase of the Alcazar, Toledo (1550–1574) was the product of Covarrubias, until he retired in 1562. It was completed by Juan de Herrera. This staircase is set on the axis of the courtyard and in its final form occupies the whole of one side. Like the other Spanish open well stairs discussed, it opens onto a cloister and is separated from it visually by the inner arcade of the cloister. The immense central bottom flight terminates in and is divided by three of the arcade segments. From the landing at the top of this flight, two flights (of about one-third the size of the bottom flight) ascend in opposite directions and at right angles to the

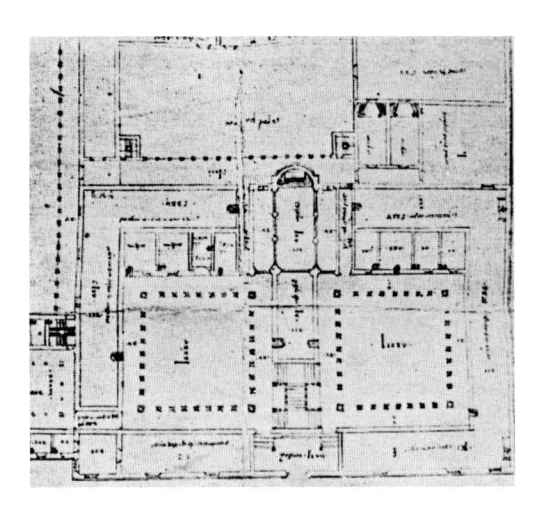

3.15 Plan of Hospital of San Juan Bautista, Toledo, reprinted from Wilkinson (1975), 68

Composite Stairs

first one. From landings at the top of these two flights, flights of the same dimensions return parallel to the first flight, ending at an arcade opening into the top cloister.

While this stair was being developed into its final form, work started on the new royal palace outside Madrid, the Escorial (1563–1584), designed by Juan Bautista de Toledo and Juan de Herrera. There can be little dispute that the grandeur of the royal palace affected the final stages of the construction of the Alcazar in Toledo, for Philip II and Herrera were the driving forces of both. The Alcazar stair (fig. 3.16), within its own spacious hall and separated from the courtyard, in feeling and scale foreshadows the baroque stairs. It is not yet an imperial stair in configuration, nor does it have the richer spatial and decorative accoutrements of even the Escorial stair. However, it is much closer to the baroque than the plateresque of the earlier rectangular open well staircases. The Escorial stair has many similarities to that of the Alcazar; it is symmetrically located on the axis of the monastery entrance and the Courtyard of the Evangelists, and it is accessed from the cloister and through the arcade enclosing it. The arcade provides the gateway to the stair and forms a layer between the stair and the cloister. The stair hall is of double height with its own vaulted ceiling and is decorated with a severe austerity. For the first time, the Escorial stair also uses the so-called imperial plan. The term is derived from the Spanish *escalera imperial,* used in sixteenth-century Spain and reintroduced by Pevsner (1968). By his definition, the imperial stair starts with one straight flight. At the landing, one turns to left or right through 180 degrees and then ascends parallel to the first

3.16 Alcazar stair, Toledo

flight. Alternatively, the stair starts with two flights and finishes with one leading up to the upper floor.

The designer of the Escorial imperial stair (fig. 3.17) is still a matter of dispute, but the influence and perhaps direct involvement of Giovanni Castello (called Il Bergamasco) is clear. Again Francesco di Giorgio's sketches may have been the source for the design. Bergamasco probably entered the service of Philip II in 1564, about a year after work started on the Escorial. He came to Spain after a successful career in Genoa in the circle of Galeazzo Alessi. His work in Genoa included the Palazzo Doria (1563) and the Palazzo Carrega-Cataldi (1558-1566). The latter contains a noteworthy T-shaped open grand stair in the entrance lobby. The Palazzo Doria-Tursi (1564), now called the Palazzo Municipio, was being completed at about the same time that Bergamasco left Genoa. It was probably designed initially by Lurago and completed with the assistance of Domenico and Giovanni Ponzello. This palazzo is of interest here because it contains a rectangular open well stair of strikingly similar layout to Covarrubias's Alcazar in Toledo and the Viso del Marques palace in southern Spain (fig. 3.18). It is also reasonable to conclude that the Genoese and Spanish architects were familiar with each other's work. Wilkinson (1975, 77) suggests that the architects for the Municipio knew of the Toledo Alcazar because their patron, Grimaldi, had spent several years in Madrid; however, the Municipio was completed several years before the final version of the Alcazar.

Bergamasco's success at the Spanish court was limited. Nevertheless, before his death five years after coming to Spain, he designed the palace of Viso del Marques at Viso. The palace plan is noticeably similar to Covarrubias's Alcazar in Toledo. There is, in fact, evidence that Bergamasco was director of works for the Alcazar for a short period in 1568 (Wilkinson 1975, 77). The staircase at Viso, as in the Alcazar, was located at one end of the central courtyard on axis with and opposite the main entrance (fig. 3.19). This follows the plan pattern of the Palazzo Municipio in Genoa. Spatially and in its decorative treatment, however, the staircase of Viso del Marques is much closer to its Italian Renaissance predecessors. The flights are vaulted and enclosed, with no side view. The emphasis is limited to the perspective views and to the rich decoration of the vaults and walls.

The design for the Escorial stair went through several stages (fig. 3.20). There are records of an early version by Juan Bautista de Toledo dating from about 1565, and there is no doubt that for several years he was involved in the design of the Courtyard of the Evangelists and the staircase. He proposed four flights around a fountain in an open stairwell with three solid surrounding walls. This design is much closer to the open rectangular wells of fifty years earlier than those of Covarrubias or Bergamasco. Bergamasco himself made a model for the staircase in about 1567, perhaps within months of the death of Juan Bautista. In 1571 the design was changed again, and this time, alterations were made in what had already been constructed. By 1571, Bergamasco was dead and Juan de Herrera seems to have taken charge, so the changes and the final form may well have been his.[2]

Composite Stairs

3.17 Imperial stair of the Escorial, near Madrid

3.18 Viso del Marques palace, Viso, Spain

From Square Newel to Imperial Stair

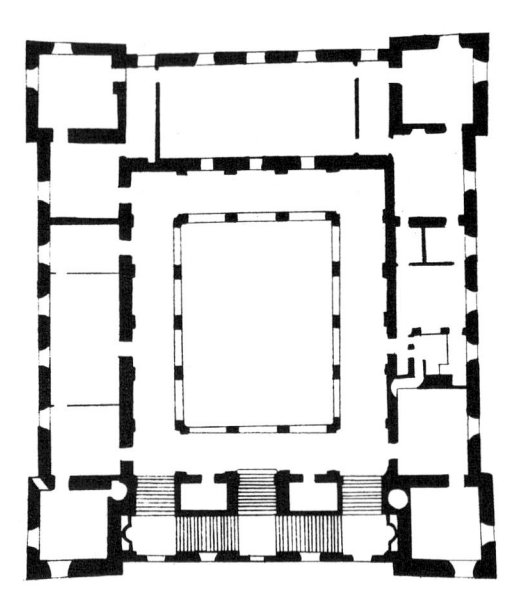

3.19 Plan of Viso del Marques palace, reprinted from Wilkinson (1975), 74

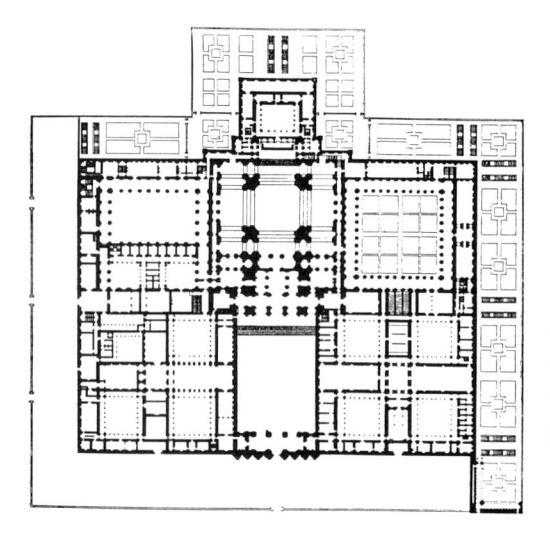

3.20 Plan of the Escorial (1589 plan by Juan de Herrera), reprinted from Wilkinson (1975), 66

3.4 The Influence of Garden Stairs

The general form of the Escorial imperial stair presaged the advent of the great baroque stairs, but it has little of the scenographic presence or spatial exuberance of those of the palaces of Bruhl, Wurzburg, Pommersfelden, or Bruchsal that are described later. Obviously, there are other significant factors besides the layout that influenced the flowering of the design of these great staircases.

There are few examples of notable internal stairs to be found in the architecture of classical Rome; most of the interior stairs are tucked away within service shafts. In gardens, sanctuaries, and fora, however, stairs as articulators of space reached a level of sophistication that was certainly part of the inspiration for the extraordinary interest in stairs during the Renaissance. Many of the spatial theories that exploit stairs in the interiors of Renaissance and baroque palaces first appeared in garden design, and clearly some of these great interior designs were intended to represent the spatial effects of the garden stairs.

The Renaissance gardens provided the inspiration. The typical large medieval garden was in effect subdivided into secluded rooms for the various types of plants—herb gardens, vegetable gardens, orchards—and the walls of these rooms were formed from hedges, fences, or even masonry. These functional subdivisions became the basis for the formal gardens of Europe, reinforced by the influence of the four-square Persian Paradise garden. By the high Renaissance, the architectonic exploitation of the garden took place with real vigor. Sloping sites with distant views were preferred, so a variety of new stair and ramp designs were developed. Or pos-

Composite Stairs

sibly the excitement in the many discoveries of the potential of stairs generated a new interest in the sloping site, the typical expression of the Italian Renaissance garden.

At the beginning of this exuberant period of garden design—at the Villa Medici at Fiesole (1458–1461), for example—we find one of the first of these terraced gardens. Although the design has moved far from the typical medieval garden, it does little to exploit the visual spatial possibilities of the hillside. We find nothing of the spectacular use of space-connecting stairs and ramps that was quickly to become such a tour de force at the hands of the Renaissance garden architects. The challenge that this new generation of garden architects set for themselves was to try to connect the horizontal planes and terraces lying on different levels into a continuous visual architectonic experience. The medieval separation of the various parts of the garden was eschewed in favor of the dramatic possibilities of linking each distinct room of the garden to its neighbors through visual devices—overlooks, stairs, vistas, accents, and water.

Donato Bramante (with Pirro Ligorio) is usually credited with creating the first of these Renaissance gardens at the Cortile del Belvedere of the Vatican (figs. 3.21 and 3.22). Bramante started the work; after his death in 1514, Ligorio took over. Bramante's task was to try to link in an architectural composition the Belvedere summer residence of the pope and the Vatican Palace, located about 330 yards away and below it. Pope Julius II wanted the whole court to be composed in the "Roman manner." Clearly imperial Roman planning provided the inspiration for the design.

Bramante's project is remarkably like that of Ligorio's drawing of a reconstruction of the Roman Temple of Fortuna at Praeneste (Palestrina) of 80 B.C. (fig. 3.23). Both designs culminate in a monumental semicircular *exedra* (sanctuary) set into a screen that opens onto a terrace. In both, this terrace is linked to the area below by a pair of dogleg stairs set symmetrically at right angles to a central axis. The focal element and the axial vista became dominant features of Renaissance and baroque garden design. The second terrace is linked to a third by another grand stair, of different pattern. The stairs were used to bind the whole into a single spatial and functional unit. The eye of the observer was led up the diagonal planes of the stairs to the terraces above. The stair became, as Giedion (1964) puts it, "the symbol of movement."

There is evidence of sophisticated use of the external grand stair by the Imperial Romans. The Roman gardens of Lucullus, for example, sited where the Trinità dei Monti now stands at the top of the Spanish Steps, was apparently approached by terraces connected by stairs, according to drawings made by Ligorio. This layout, composed of terraces at right angles to a dominant axis and connected by a series of stairs and ramps, was the quintessential exemplar of what were to become the basic precepts of Renaissance garden architecture. An even more obvious reference to the Lucullan gardens and Praeneste can be seen in Ligorio's design for the Villa d'Este at Tivoli (1550–1580s). Here, water and stairs set the stage for a landscape of drama that is unsurpassed.[3]

The use of water with stairs is typical of these Renaissance gardens. The architects were quick to see the architectonic parallels

3.21 Sketch of the Belvedere court, Vatican

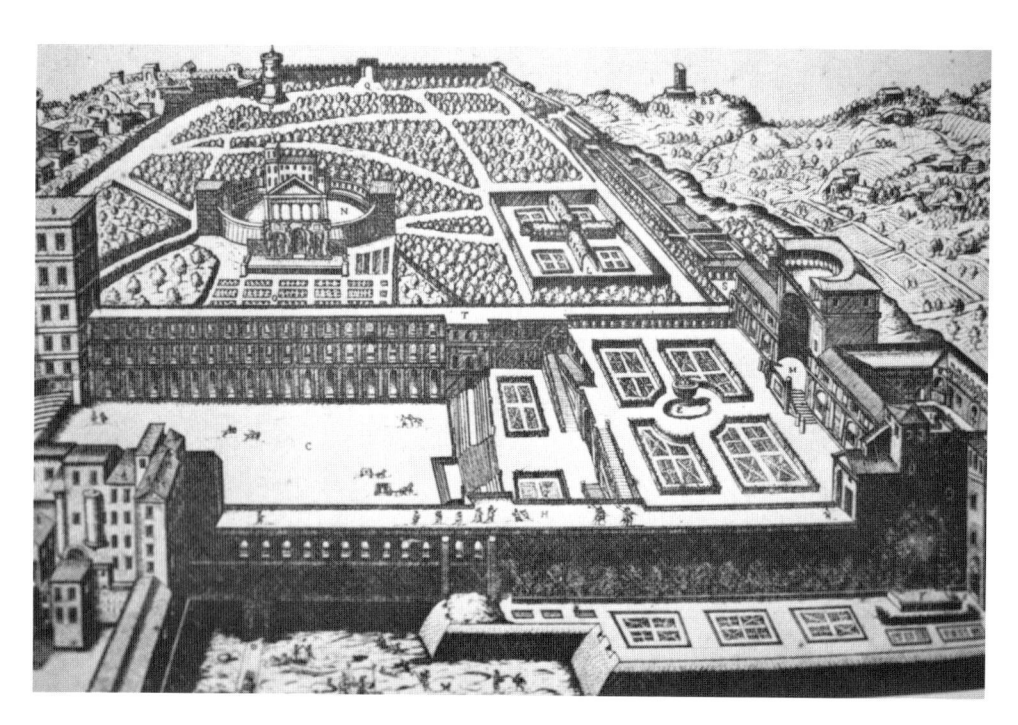

3.22 View of the Belvedere court, Vatican

Composite Stairs

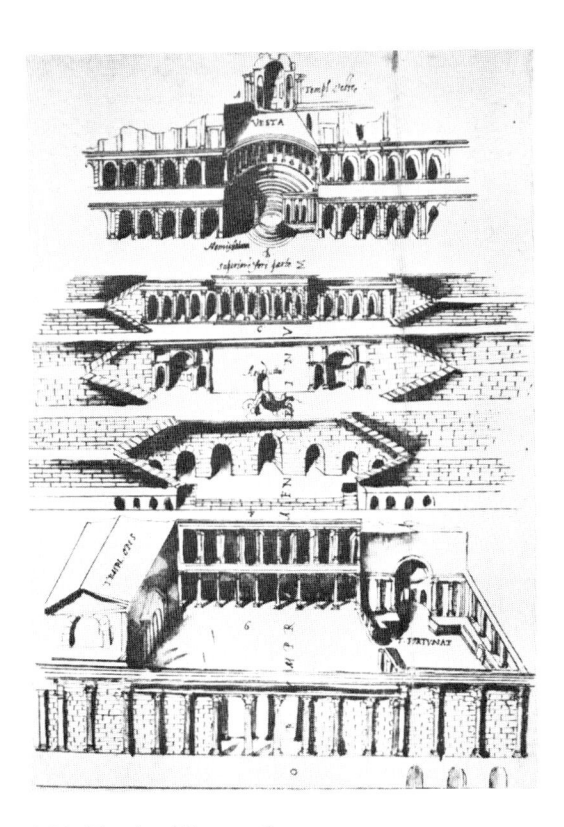

3.23 Temple of Fortuna, Praeneste

between natural cascades of water and the way the flow of space down a hill can be suggested and controlled by the use of stairs and ramps. The next creative leap was to combine water and stairs to exploit the sensory and architectonic possibilities of both. Ligorio's garden at the Villa d'Este makes full use of fountains, water, stairs, plant materials, and ramps to form this Renaissance tour de force. It is laid out with the intention that one should enter at the bottom of the hill (rather than at the top as one does today) and then follow one of the several gently rising paths, stepped ramps, and stairs, each offering an intriguing vista, up to the many terraces. Eventually, after experiencing some of the most sublime and astonishing visual effects, one is exposed to a panoramic view from the top, where the villa dominates the countryside spread out below. The terminal and junction points of the ramps and stairs of the Villa d'Este are punctuated by fountains. The stairs are still simple flights without the sculptural intricacies that were to follow soon; however, the spatial ambience of the high Renaissance and baroque stairs are already present. The stairs are no longer simple utilitarian connectors; they are architectural events to be seen from a distance and then experienced as a changing sensory sequence.

Giacomo da Vignola's contribution to the development of garden design and stairs matched Ligorio's. The Villa di Papa Giulio in Rome (probably designed with Vasari and Ammanati in 1550) and the Villa Farnese at Caprarola (1547–1587) contain some extraordinary examples of garden stairs. In the Villa di Papa Giulio, the Renaissance genius for treating landscaped changes of level was added to the growing preoccupation with methods for increasing the interpenetration of house and garden. The house, in fact, provides a framework for enclosing the outside spaces so that they are formed into a series of linked garden rooms. A nympheum is used to connect the main court to a small formal garden that lies at a lower level. Two segmented flights of stairs carry both the eye and people up to a porticoed screen separating the two parts of the garden (fig. 3.24). The two arms of the stairs embrace and enclose the formal garden architecture of the nympheum like the wings of a stage. This resembles Ligorio's stairs at the Villa d'Este leading to the topmost terrace.

At Caprarola, Vignola uses a gentle stepped ramp framed between two walls to connect the casino of the secret garden to a copse of trees. The two enclosing walls repeat the walls formed by a vista cut through the trees (fig. 3.25). The gentle walking movement suggested by the stairs is accompanied by a rippling cascade water stair, which divides the pedestrian's stair into two parts. Vignola uses the same idea at Villa Lante (fig. 3.26).

The flow of the water at Caprarola is slow and quiet. In effect, one walks beside a languid, meandering stream that murmurs softly. Because the slope is gradual, movement is exaggerated by shaping the edge of the channel to form volutes (in the shape of fishes) to cause, as well as to suggest, turbulence in the water in much the same way that rocks on the edge of a

3.24 Nympheum of the Villa di Papa Giulio, Rome

3.25 Garden stair, Villa Farnese, Caprarola

3.26 Runnels at Villa Lante, Bagnaia, Italy

The Influence of Garden Stairs

stream might. The stair provides an axial approach to a small amphitheater that serves as an antechamber for the casino. It connects the casino to the pool with its single plume fountain and to the vista beyond. It links the woods to the house visually and functionally, gradually merging the natural landscape to the carefully sculptured formal treatment of the buildings.

Larger and steeper stairs suggest more forceful rapids, so we find the stairs in the Villa Corsini at Mezzomonte flowing as it were in a ravine and sharing the space with a mountain stream falling down giant steps (fig. 3.27). Again there is an attempt to integrate nature and the stair, at least symbolically, and to reinforce the metaphor of stair as a waterfall. Later, as baroque monumentality developed competitively, these cascades increased in scale until we find those at the Royal Palace at Caserta dominating the grand vista (fig. 3.28).

One of the most ingenious of these great theatrical landscape water stairs is that designed at Frascati (1598–1604) for Cardinal Aldobrandini by Giacomo della Porta and Carlo Maderno. The water for this continuous performance (see Falda's view as one would see it from the upper part of the villa, fig. 3.29) starts in a grotto, from which it flows down the central axis in a variety of cascades, falls, and rills between a pair of stair flights. Ultimately the stream reaches the water theater—a lofty, semicircular screen of niches and statuary alive with the play of water. Stairs curve down around the outer flanks of the water theater, and here the stair users become active participants in the drama. Charles de Brosses described in 1739 (quoted by Thacker 1979, 114) how "as soon as you

3.27 Garden stair, Villa Corsini, Mezzomonte, Italy

3.28 Water stair, Caserta, Italy

3.29 Falda's sketch of the stair at Villa Aldobrandini, Frascati, Italy

The Influence of Garden Stairs

are part of the way up, the water-jets shoot out, crisscrossing in every direction, from above, below and the sides." (See Falda's 1675 view, fig. 3.30.)

Perhaps the merriest of these water participant stairs can be found in the sunken, secret garden of Flora at the Villa Torrigiani near Siena built in the second half of the seventeenth century. Georgina Masson (112) says that the sunken garden

is connected to the lemon pool garden by double staircases which, although perfectly scaled to the diminutive size of the garden are carried out in the very grand manner indeed that probably no other country or period could have equalled. But as the grand ladies and their cisisbei, clad in their seventeenth century silks and satins, descended these stairs to walk among the flowers and scented herbs of the parterres, their host could imprison them in his secret garden by raising a veritable wall of spray from fountains concealed in their highest steps. As they ran forward, seeking some way of escape, by turning a tap he could pursue them down the whole length of the garden with surprise showers hidden among the pebble mosaics of the paths. When at last his guests thought they had found a refuge in the little temple of Flora at the far end, more deluges awaited them, forcing them to climb the stairs to the terrace above, only to receive a final soaking from a flower-garlanded statue of "Flora" herself.

These gardens, with all their dei ex machina, are intended as a backdrop for the theater—the pageant of life in these great outdoor rooms, where the players' desires for ceremony, drama, surprise, excitement, seclusion, or solitude are provided with the appropriate setting.

The development of the garden as architecture owes much to the contribution of

3.30 *Water theater, Villa Aldobrandini*

Composite Stairs

garden stairs. In fact, in many of these great Renaissance and baroque gardens, stairs were made into the main element, central focus, or principal architectural event. The Villa Garzoni at Collodi (second half of the seventeenth century) and the Villa Bettoni (1760) at Bogliaco are notable examples. Stairs form the focus of both gardens. At Villa Bettoni, four matched pairs of stepped ramps (fig. 3.31) ascend to successive terraces. The ramps start on each side of an arched grotto and terminate at a dummy facade with niches. From this level, they ascend again to another grotto, up again to facades that form the end of trellises, and then on once again to the top level and the focal vista. Villa Garzoni is somewhat more modest but quite similar in layout (fig. 3.32). Each pair of stairs leads to a terrace and climbs about a deep arched niche. The influence of these great baroque garden stair compositions continues; Ricardo Bofill pays homage to them in his recent Les Echelles du Baroque (Ladders of the Baroque) in Montpellier (fig. 3.33).

The garden stair is used in these villas for a controlled journey through the landscape. The variety and richness of these experiences was later to be generated artificially and theatrically in the great baroque stairs. The drama of the outdoors was staged in an idealized manner indoors with the sky frescoed onto the ceilings and the movement of the landscape suggested in marble or painted onto the walls.

3.31 *Villa Bettoni, Bogliaco, Italy*

The Influence of Garden Stairs

3.32 Villa Garzoni, Collodi, Italy

3.33 Ladders of the Baroque, Montpellier, France, by Ricardo Bofill

Composite Stairs

3.5 Learning from Formal Entry Stairs

Garden stairs enjoy considerable freedom of composition, and the Renaissance architects were quick to explore the rich, dramatic possibilities. The stair was given many forms and was used in many ways and quite frequently was the architectural event around which the garden was composed. Nevertheless, stairs as parts of buildings are by nature subsumed by the whole and therefore constrained by the design of the remainder. During the baroque, stairs became the principal ornament of the great palaces of the day. Even the greatest of these staircases of honor, however, struggled to compete with the natural drama and dramatic effects of light, space, transition, water, and planting that were typical of garden stairs.

The Formal Approach

If garden stairs became the model and inspiration (albeit transmuted) for the great baroque interior stairs, stairs that formed the formal approaches to buildings were the first demonstration of the possibilities of the stair as baroque architectural theater constructed as a building element. In the baroque staircases of honor, it is apparent that the formal exterior approach stairs appeared earlier than the interior examples, using the lessons learned from the new drama of the garden stairs. The garden stair had to be tamed and understood as an element of the facade before it could be considered for inside use.

Crepidoma and simple straight flight exterior stairs were usual features of Roman temples and other buildings. The grand exterior freestanding approach stair (but not the podium) was rare until quite late in the Renaissance. The Scala dei Giganti of the Doge's Palace (figs. 3.34, 3.35, and 3.36) in Venice (1489–c.1501) is an early exception; however, it is set within the grand cortile rather than on the external facade. The stair was, in fact, designed to be the stage setting for the finale of the coronation of a new doge, Leonardo Loredan (Sohm 1985, 127). This sort of political theater was a commonplace activity on the broad stairs extending in front of some medieval town halls, which served as stages for public events and shows.[4] In Venice, the coronation procession began at the basilica, wound around the piazza, and entered the cortile through the main entrance to the palace, the Porta della Carat. The ceremony culminated with the investiture on a rostrum at the top of the great stair. The finale could thus be seen by the crowds gathered in the cortile (fig. 3.37). During the early part of the sixteenth century, it became customary for the doge, seated on a chair at the top of the stairs, to receive guests. They would have to ascend to this monumental throne under the gaze of the doge and his councilors. After the completion of the Scala d'Oro around 1560, this stair became the new locus for reception. The doge awaited his visitors on the first landing or, if the visitor was of elevated rank, at the bottom of the stair.

Designed by Antonio Rizzo, the Scala dei Giganti forms the visual terminus of the Porta della Carat, located on the axis of the gallery approach that connects the Campanile of the Piazza di San Marco to the cortile.[5] From the piazza, one passes through the shaded vaults of the Portega Foscari, out through the arch into the bright light of the cortile, and up the great stair, an extraordinary architectural sequence. The stair, flanked and dominated

3.34 Scala dei Giganti, Venice, view from the Porta della Carat

3.35 Scala dei Giganti, view from the cortile

114

Composite Stairs

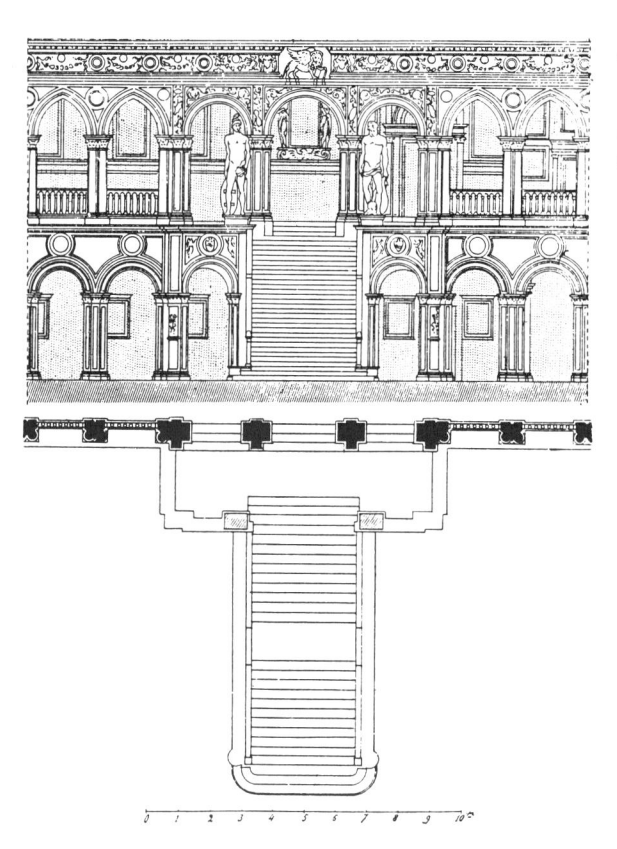

3.36 Plan and elevation of the Scala dei Giganti

3.37 Gabriele Bella, Investiture of the Doge, reprinted from McAndrew (1980), 96

Learning from Formal Entry Stairs

by Sansovino's huge figures of Mars and Neptune, is the dominant feature of the cortile, providing a diagonal accent to the irregular courtyard. The stair is set into a leg and defines the south side. In plan, the stair appears to sit in a sort of external stair hall; actually the smaller court is visually part of the grand cortile.

Organizing the Stair into the Symmetrical Facade

Stair design as a usual and integral part of the main facade of buildings had to await the exemplar of Michelangelo and his design for the Palazzo Senatorio for the Capitoline Hill (1538–1578). In this one building program, Michelangelo stated and resolved many of the contemporary architectural conundrums: creating a formal entrance to the piano nobile from the outside, organizing a symmetrical facade to include large diagonal elements, developing a new way of mediating between inside and outside spatially, and exploiting the stair as a spatial experience in relationship to the building and the piazza. Architectural theory of the high Renaissance and baroque assumed the palace with a piano nobile and a balanced symmetrical facade. The problem that Michelangelo faced—how to provide an appropriate architectural connection between the formal approach to the palace and the piano nobile—became one of the central architectural questions of the age. Architects delighted in devising external and, later, internal stairways of ever-increasing grandeur and drama to make this connection. Often the stair itself dominated the composition.

Before Michelangelo's facade for the Palazzo Senatorio, external grand approach stairs were seldom used. When they were, they were usually set at right angles to the facade, as in the Scala dei Giganti of the Doge's Palace. To form a diagonal in front of a symmetrical facade would be to set up tension that could not be resolved, even by a dedicated mannerist, without the introduction of some balancing element, so Michelangelo invented, reinvented, or developed the double or paired stair (fig. 3.38), where a second, matching flight is used to balance the first in the composition.

Such an arrangement had been used before, although probably not as a major element in a symmetrical facade. The amphitheater at Pompeii (A.D. 59) (figs. 3.39 and 3.40) has paired flights, and Bramante and Ligorio used this device twenty-four years earlier in 1514 to form a symmetrical approach along the main axis within the Belvedere court (inspired by the Temple of Fortuna at Praeneste). Michelangelo was familiar with the garden stairs of the Belvedere and in fact returned the compliment by using the pattern of the Senatore stair for a remodeling of Bramante's exedra in 1551.

Leonardo may also have been a source. Although he died some nineteen years before construction started at the Campidoglio, his sketches for a domed and centrally planned church show exactly the same arrangement of stairs with a porticoed entrance (fig. 3.41) (Vollmer n.d.). Interestingly, Michelangelo's original sketches for the Campidoglio show a similar portico that was not built (fig. 3.42). Furthermore, Leonardo had worked closely with Bramante in Milan in 1496, some years before Bramante had started work on the Belvedere, and clearly Bramante was familiar with and influenced by Leonardo's sketches of centrally planned churches.

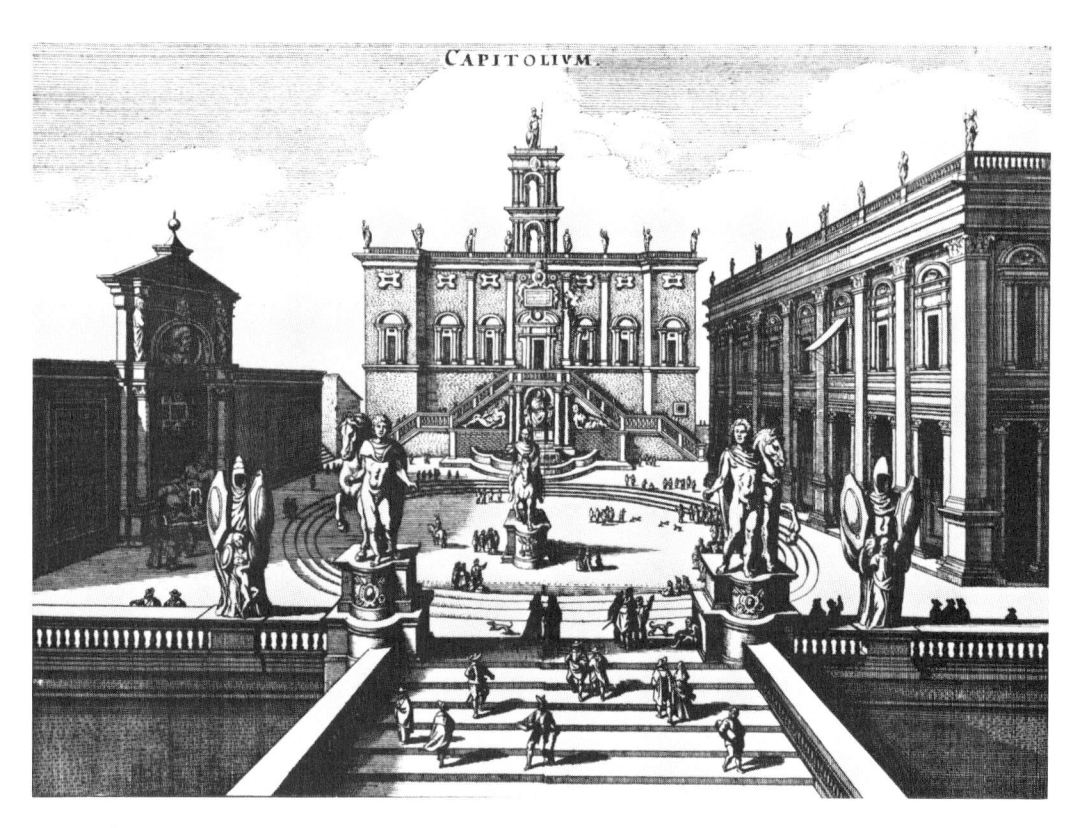

3.38 Palazzo Senatorio, Rome

3.39 Mural depicting the riot in the amphitheater, Pompeii, reprinted from Picard (1968), 66

Learning from Formal Entry Stairs

3.40 *The amphitheater, Pompeii*

Composite Stairs

The dominance of the stair as part of the Campidoglio composition is reminiscent of the vestibule for the Laurentian Library. Michelangelo did not conceive of the Campidoglio as a composition of three blocks but as a contained space, like a three-sided room with no roof. It is a reception room for the palazzi and a room for open air public ceremonies.

3.6 THE RISE OF THE INTERNAL GRAND STAIR

In medieval Europe the stair as an access way was first employed to make complex circulation patterns possible. The stair had, of course, been used since earliest times to gain access to groups of rooms. In the Middle Ages, it was used to satisfy the growing desire for privacy, to avoid crossing one room to gain access to another, to obtain separate communication between apartments or suites on different floors, or for secrecy. Corridors were unfashionable

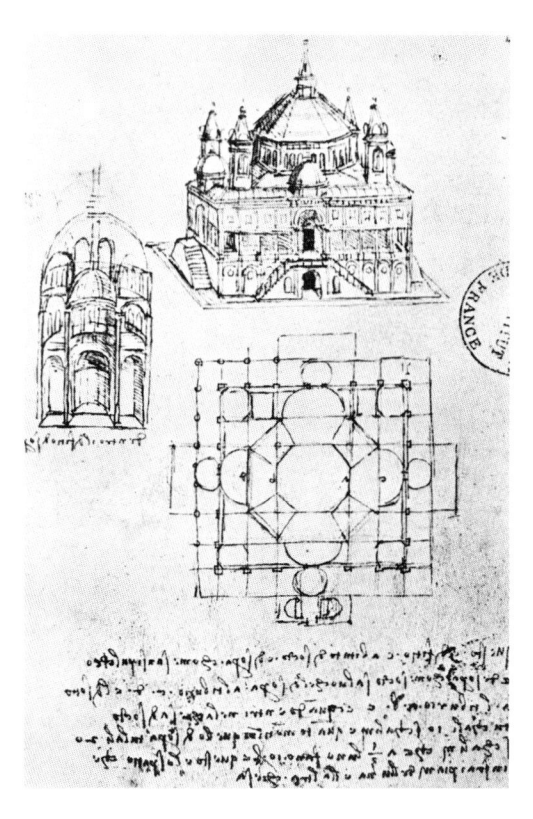

3.41 Leonardo's sketch for a centrally planned church

3.42 Michelangelo's sketch of the Senatore

The Rise of the Internal Grand Stair

or eschewed, and so the number of staircases multiplied in proportion to the number of apartments. For example, the mansion of Jacques Coeur, the banker who provided much of the money used to drive the English out of France, erected at Bourges (c. 1450) features eight staircases (fig. 3.43), and the palace of Chambord (fig. 3.44) contains something like twenty-five.

The relocation of the principal rooms of noble houses and palaces from their traditional place on the ground floor to the piano nobile made the access stair of the Renaissance and baroque of far greater functional importance than before, so it is not surprising that the staircase grew in size and magnificence. The risers decreased in height and the treads increased in depth to respond to the increased functional demands of ceremony and traffic volume.

It was with the internal grand stair of the Renaissance and baroque that the symbolic, ceremonial, and architectural possibilities of stairs came to be exploited with such consummate skill. Internal grand stairs were no novelty to the classical world. A vast full-width flight of stairs sweeps up from the cella, where the image of the god stood, to the cult image at the western end of the Roman temple of Bacchus at Baalbek. However, where two adjacent floor levels were to be connected, even in palaces, the stair was, by comparison to the extravagant baroque examples, modest and unambitious. This is not altogether surprising, for, with some notable exceptions (Diocletian's palace at Spalato had the chief rooms of the palace on the second floor), the most important rooms and reception areas of Roman palaces were situated on the ground floor (Robertson 1969, 319).

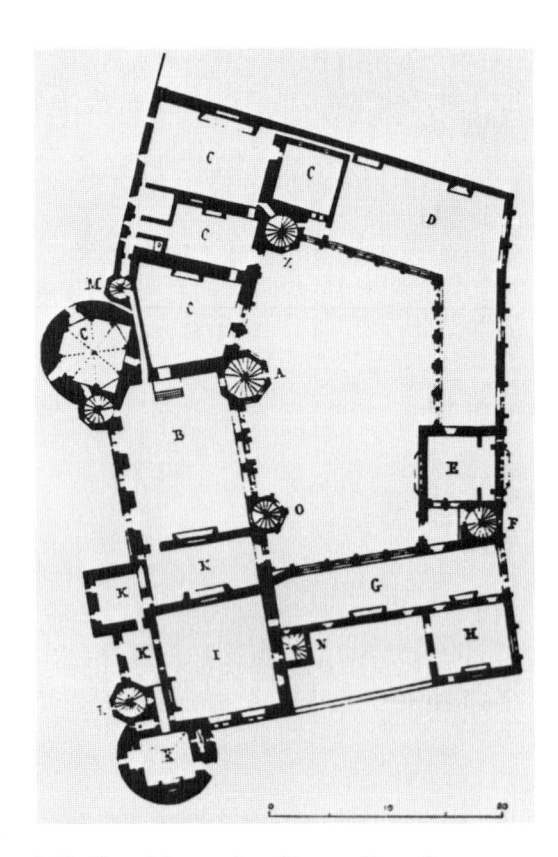

3.43 Plan of the mansion of Jacques Coeur, Bourges, France

Composite Stairs

3.44 *Palace of Chambord, France*

The Rise of the Internal Grand Stair

Pevsner (1968, 278) brushes these Roman stairs aside as being only "vaulted corridors running up at an angle." And Guadet (n.d., 412), rather slightingly, points out that the Latin language had but a single word, *scala,* to express ladders and stairs (and in truth Roman stairs were usually steep).

The grand staircase in its hall was a typical high Renaissance and baroque solution to an architectural conundrum or an architectural anomaly: how best to treat diagonal planes within an orthogonal grid of spaces and how to ensure the cubic volumetric integrity of spaces adjoining the stair. The initial solution was to set the stair within its own distinct container connected to, but quite separate from, adjacent volumes. Michelangelo's Laurentian Library is a superb example of this. The solution presented by the separated stairhouse was inherently static in terms of spatial flow, as its designers intended. One entered the stairhouse and was entertained by its theatrical devices to draw one's attention to the stair and to the top and bottom. One then left the container and passed into the next spatial unit in the sequence.

It was not long before architects found less confining solutions, and it is interesting to speculate whether these solutions were initially developed in garden architecture, for, as we have seen, quite early in the Renaissance architects quickly mastered the means for establishing and connecting the parts of their landscaped gardens by the shrewd manipulation of stairs and ramps. These gardens still contained the basic elements of Renaissance theory in that each space was made to be aesthetically discrete unto itself. The theory, however, had to be tempered by the reality of the limitations of plant materials and a ceiling formed by the sky. A degree of spatial continuity was inevitable and eventually was welcomed and sought after in designs. This occurred when architects realized that adjacent spaces could be compared and contrasted with each other more dramatically by developing these spatial linkages.

Inside buildings, the first transference of these ideas occurred in Spain. The next developments are found within the hillside palaces of Genoa, which were set within and incorporated part of terraced gardens to a remarkable degree. Like the Villa d'Este, they were intended to be experienced sequentially, starting from their main entrances at the level of the street.

The sequences of spaces and the way these are connected by stairs are well demonstrated by several palaces. In the Palazzo Municipio (1564), for example, the parallels to the Renaissance theories demonstrated at the Cortile del Belvedere and the Villa d'Este (on which construction started fourteen years earlier) are obvious.[6] One enters the palace from the street into a shady vestibule (like that of Palazzo Marcello-Durazzo) dominated by a straight flight stair. Attention is immediately drawn to the open court at the top.

In the Palazzo dell'Università by Bartolomeo Bianco (1623), yet another of this group of Genoese palaces, the vestibule stair is guarded on each side by a pair of lions struggling to avoid slipping down the stairs (figs. 3.45 and 3.46), and the flight is tapered to form a false perspective to magnify its length and grandeur (fig. 3.47). The flow of space in the Università is inexorable and singular. The whole wall at the top of the stair is removed, and the vestibule and the courtyard, despite their different levels, are interconnected through a vaulted colonnade.

Composite Stairs

3.45 Palazzo dell'Università, Genoa

3.46 Lion Balustrade, Palazzo dell'Università

The Rise of the Internal Grand Stair

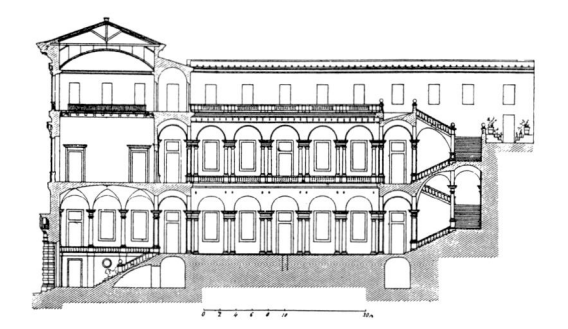

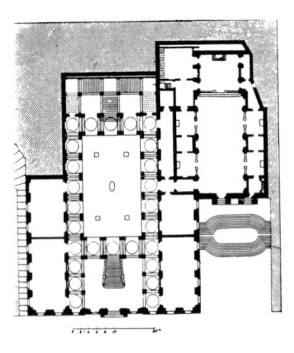

3.47 Plan of the Università

In both palaces, from the cortile one's attention is attracted to the only diagonal element—the stair at the opposite end of the axis—and to the upper floor colonnade above it, which frames a glimpse of the garden on the upper terrace. This stair itself is a new layout, perhaps based on Francesco di Giorgio's sketch for such a stair for a Palazzo della Repubblica. The first flight of this stair is set centrally on the same axis as the stair leading from the lower vestibule. At the landing, the stair divides and forms two flights at right angles to the first. After reaching two more landings, two final flights parallel to the first lead to the vaulted colonnade that surrounds the cortile at the piano nobile to form a sort of prototypical imperial stair (fig. 3.48). The view looking back from the top is over the stair through the open loggia and into the garden terrace behind. The stair and garden are integrated once again.

To emphasize the stair and to carry the eye even higher, the plane of the stair, without steps but with the balustrading, is carried up to the roof, forming a purely architectonic, nonfunctional continuation of the stair (fig. 3.49), a trick adopted by James Stirling at his Stuttgart art gallery (fig. 3.50) and also used by Bernard Tschumi in one of his follies in La Villette park in Paris, where the helical stair turns out to be a watercourse, not a stair (fig. 3.51).

During the baroque, by contrast, with the connection to the piano nobile now so important, the status of the building and its owner was expressed as much by the stair as by the facade, the portico, and the reception rooms. The amount of space used to house the stair grew to immense dimen-

Composite Stairs

3.48 Università stair from the courtyard

3.49 Università false stair

The Rise of the Internal Grand Stair

3.50 False ramps, Stuttgart art gallery

Composite Stairs

3.51 Fake spiral stair in La Villette park, Paris, by Bernard Tschumi

3.52 Stair at Caserta, by Luigi Vanvitelli

sions. By no means the largest is Neumann's grand stair at the Wurzburg palace, which occupies about 102 by 64 feet (31 by 20 m). Luigi Vanvitelli's stair for the Royal Palace at Caserta (Wittkower 1966 describes the palace as "the overwhelmingly impressive swan song of the Italian Baroque") (fig. 3.52) occupies some 13,855 square feet (1287 sq m), being 163 by 85 feet (50 by 26 m). The oval staircase hall at the Episcopal Palace at Bruchsal is, significantly, the largest room in the palace.

It was not, of course, sheer size that was the goal. The stair had to express the taste and elegance of the aristocracy. It also had to be an elaborate backdrop, rather like a stage set, for ceremonial occasions, with treads generous enough and risers shallow enough for ladies to walk not only with safety but with poise and dignity. Stage design undoubtedly influenced these extraordinary baroque stair fantasies.

The Rise of the Internal Grand Stair

Above all, these grand internal stairs—Staircases of Honor, as Guadet (n.d.) calls them—were designed to make the transition from the ground to the piano nobile and the upper floors as imperceptible a spatial barrier as possible. The stairs and the stair halls are used to distract the stair user from the act of climbing and to integrate the vertically adjacent spaces.

3.7 THE BAROQUE STAIRS OF HONOR

The history of the great baroque stairs is the history of a coterie of remarkably talented architects. They and their patrons shared a passion for perhaps the most theatrical architectural building element in an age strongly influenced by the fantasies of theater set designers. The movement gathered momentum in the second half of the seventeenth century and was over by the middle of the eighteenth; it started in Italy and flowered in Germany and Austria.

The prototypical Genoese imperial stairs in the Municipio (1564) and the Università (1623) have the same basic layout as the Alcazar, Toledo, but are visually and actually set at the end of an axial entrance from the street at the far end of a courtyard. Soon after they were completed, Baldassare Longhena built a stair of the same pattern for San Giorgio Maggiore in Venice (1643–1645). The Longhena stair is the first of the great Italian monumental internal stair halls, much grander than its Genoese forebears (fig. 3.53). As Wittkower (1986, 300) points out, this staircase hall prefigures the extraordinary staircases that developed later north of the Alps and prepares the way for the scenographic architecture of the eighteenth century.

The San Giorgio Maggiore stair is the lineal descendant of Michelangelo's stair and vestibule for the Laurentian Library of a hundred years earlier. Like it, Longhena's stair sits wholly within the stair hall, not attached to a cloister or court. One enters the stair hall before reaching the first flight, experiencing the space as a grand stair within an immense room brightly illuminated by windows on three sides at the upper level. The hall is roofed with a vault that is clearly articulated from the enclosing walls as if anticipating the scenographic ceilings of the baroque stair halls with their fantasies set in an artificial sky. The stair hall, however, is treated with decorative restraint and is more like a spacious hallway than a reception room.

About twenty-five years after the completion of San Giorgio, Louis Le Vau and Charles Le Brun designed the Escalier des Ambassadeurs (fig. 3.54) for the palace of Versailles. The stairway was intended for the use of ambassadors ascending to the Hall of Mirrors for an audience with Louis XIV. This stair superficially resembled Longhena's stair in layout, but the theatrical explorations so typical of baroque stairs had now begun to appear. The bottom flight flowed out from the landing in all directions. Illusionary scenes filled false windows and niches, and the whole was decorated with marble inlays and gilt bronze, painting, and sculpture. Trompe l'oeil effects with false perspective were used to expand the space, which was lit from above by a skylight. In 1752, less than a hundred years after it was built, Louis XV destroyed the stair to build new apartments for his daughter. Our knowledge of the stair has been amplified through a scale model constructed in 1958 based on engravings and extant decorative fragments (*Connaissance des Arts* 1958, 70–77).

Composite Stairs

3.53 San Giorgio Maggiore, Venice

3.54 Escalier des Ambassadeurs, Versailles

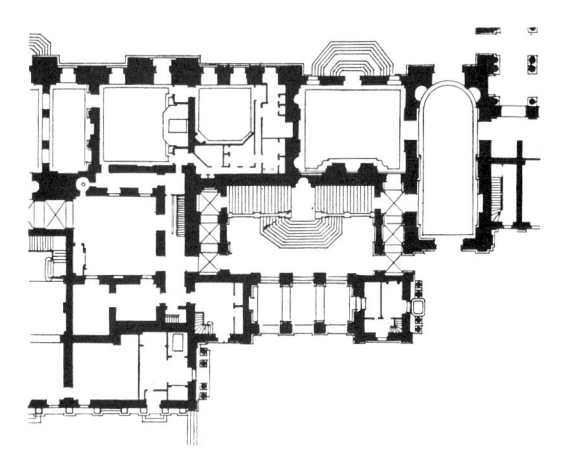

The Baroque Stairs of Honor

It was in Germany and Austria that we find the grandest and richest of the great baroque stairs—at the town palace of Prince Eugene by Fischer von Erlach; at Pommersfelden, Schloss Mirabell, the palaces of Daun-Kinsky, and the Belvedere by Lucas von Hildebrandt; and at the episcopal palaces at Bruchsal, Bruhl, and Wurzburg by Balthazar Neumann.[7] These stairs cannot be considered simply as sumptuous means to pass from the ground to the reception rooms. They became an end in themselves—resplendent art objects, the jewel in the crown of the palace. They exemplified perfectly the current theory that allowed each building element to display its own expression, even if this tended to isolate it from its surroundings. Surrounding walls were dissolved by using large windows, recesses, mirrors, sculpture, and illusionary devices with domelike ceilings transformed into artificial skies.

On these stairs, the formalities of reception and departure were played out within the diplomatic protocol of the court. Visiting dignitaries were often received on the stair, and where they were received was a measure of their rank and social position. The gentle riser-tread geometry mandated a stately, comfortable, and ceremonial gait for the procession of ascent or descent. (There were also secondary and servants' stairs for service and secret stairs for connecting manorial rooms.)

Fischer von Erlach

Johann Bernhard Fischer von Erlach (1656–1723), the oldest of this talented trio, was the son of a sculptor from Graz. He trained in Rome under an assistant of Bernini and returned from Italy after almost sixteen years, immediately making use of his knowledge of Roman architecture to impress the Viennese aristocracy. Palaces, churches, and works for the emperor followed.

His staircase for Prince Eugene's town palace in Vienna (1699) combines a sculptural sensuousness with a powerful architectonic manipulation (fig. 3.55). The stair starts with a single flight. This bifurcates at the first landing into two flights at right angles to the first flight, and these come together again in a balustraded landing over the beginning of the first flight. The space above expands as one ascends in a way that was to become the mark of the great baroque stairs. Instead of columns to support the landings, he used heroic figures—Atlas and Hercules—and for balustrading, he made wavelike serpentine volutes supporting vases, a device echoed in the works of Hildebrandt.

Hildebrandt

Johann Lucas von Hildebrandt (1668–1745) was born in Genoa, the son of a German-born officer in the Genoese army. He trained as a military engineer and was attached to Prince Eugene's army as a fortifications engineer from 1695 to 1696. After the Austrian defeat of the Ottoman Turks in 1696, Hildebrandt went to Vienna, where a flourishing program of palace building was in progress. He was an immediate success in attracting clients who wanted to improve the appearance of some existing or proposed building. His most notable stairs were for the palaces of Pommersfelden (1713–1716), Daun-Kinsky (1713–1716), Belvedere (1721–1722), and Mirabell (1721–1727).

The stair at Schloss Weisenstein, Pommersfelden (figs. 3.56, 3.57, and 3.58), sits

3.55 Prince Eugene's town palace, Vienna

within a three-story colonnaded enclosed courtyard with galleries at each level, surmounted by a great frescoed and vaulted ceiling—a sort of romanticized interior version of a Renaissance courtyard garden where the goings and comings on the stair can be observed as if from the boxes of a theater. The view on entering the corps de logis is into a court formed by the enclosing flights of the stair and through an arcade that supports the top two flights and the landing; beyond the arcade is a large niche. The great stair begins with a pair of flights set at right angles to the central axis on which one enters. After landings, the next two flights rise parallel to the axis and away from the entrance. After further landings, the last two flights rise toward the axis, running parallel to the first two, and terminate at a landing leading to the vestibule. Lothar Franz von Schonborn's influence on the design must also be recognized, as must that of the original architect, Dientzenhofer.

Hildebrandt's stair for the Daun-Kinsky Palace was constructed in Vienna for the absent viceroy of Naples at about the same time as Pommersfelden. The palace is situated on a narrow urban lot and is formed around two courtyards. Abutting one of the long walls of the first long court is a corridor leading to the rear quarters. The stair adjoins the corridor, separated from it by an arcade, and borrows its light at the ground floor from the windows onto the courtyard. The stair rises up to the next floor in a straight flight broken by an intermediate landing. The plan at this floor is the same, so to mount the next flight, leading to the principal rooms of the palace, one must pass along the corridor to a position above the beginning of the lowest

The Baroque Stairs of Honor

3.56 Schloss Weisenstein, Pommersfelden

3.57 View of Schloss Weisenstein, Pommersfelden, reprinted from Mielke (1966), 223

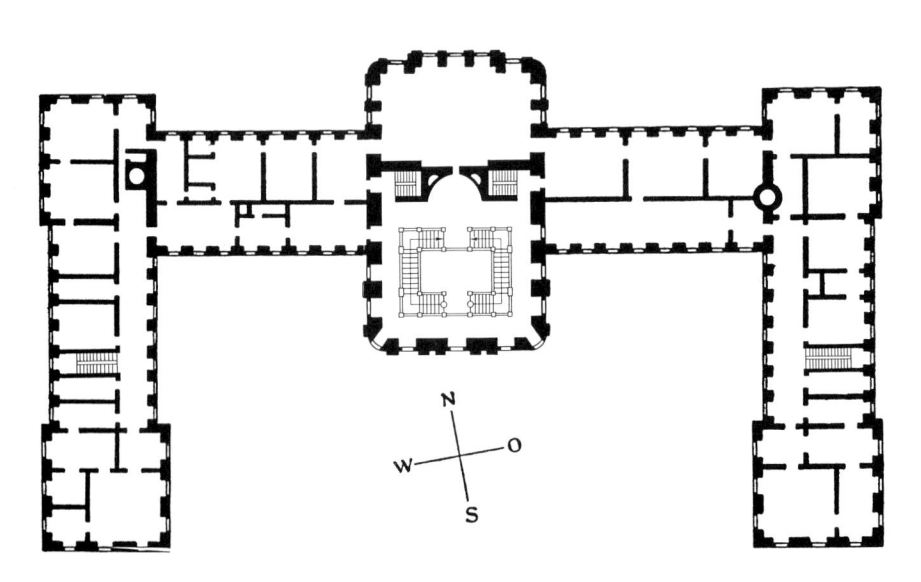

3.58 Plan of Schloss Weisenstein, Pommersfelden

Composite Stairs

flight. At the top, the corridor and stairway occupy the same space and are covered by a vaulted ceiling and a continuous gallery running around the periphery of the ceiling. The whole is brightly lit from high windows around the space.

There are no handrails, and the balustrade of the stair and the guardrail at the top take their cue from Fischer von Erlach's town palace for Prince Eugene by filling the space with scrollwork. Hildebrandt uses the same signature in his remodeling of the stair of the Mirabell Palace in Salzburg. The work was executed for Franz Anton von Harrach, the archbishop of Salzburg. He retained the well surrounded by flights on three sides but lightened it and opened it up into a single space. The frothy scrollwork balustrades are ridden by cherubs (fig. 3.59) cavorting and tumbling off marble waves, which break and cascade down the staircase, giving an exuberant movement to the old stair.

Hildebrandt's major work was the Belvedere Garden Palace (figs. 3.60 and 3.61). Its grand stair does not have the playful charm of Mirabell and Daun-Kinsky because the Upper Palace was intended only for official receptions; the Lower Belvedere served as the residence. The main entrance to the palace, set at the top of a flight of stairs (and vehicular ramps), is on the axis of the large pool and the main gate. One enters into the entrance hall, the Sala Terrena, and the lowest flight of the imperial stair is at the far end of the hall. Originally the staircase was closed off by a wrought iron gate.

The ceiling of the Sala Terrena is vaulted and supported by four Titans acting as columns (fig. 3.62). The ceiling and the walls

3.59 Schloss Mirabell, Salzburg

3.60 View of Belvedere Garden Palace, Vienna

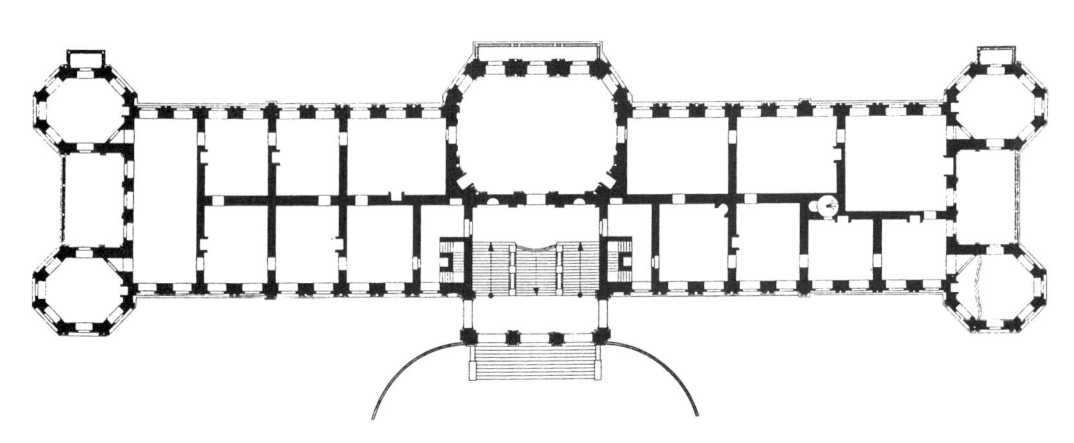

3.61 Plan of the Belvedere Palace, reprinted from Mielke (1966), 230

are decorated with allegorical stucco reliefs. The dominant central relief panel in the ceiling portrays a female figure in the sky indicating magnificence, magnanimity, and nobility.

As one ascends the staircase of honor, the view through the arcade at the landing across the formal garden to the Lower Belvedere is gradually revealed. At the top of the flight is a large rectangular hall. On both sides of the flight just ascended are flights leading to a landing and balcony that overlook the whole space and out through the external arcade to the gardens (the five arched openings were not glazed until the nineteenth century). The stair and the balcony are centered on the axis. From this balcony, one enters the Marble Hall of the piano nobile through a door in the center.

In this hall, Hildebrandt forsook the typical baroque painted fresco ceiling in favor of elaborately decorated sail vaults carried on atlantes and pilasters. In a rosette in the center of the ceiling, Alexander cuts the Gordian Knot. On the walls are reliefs portraying Greek mythological scenes. The stair is embellished with massive newel posts surmounted by lanterns carried by groups of cherubs.

BALTHAZAR NEUMANN

Balthazar Neumann (1687–1753), the third of this group of baroque stair designers, was born in Bohemia, the son of a clothmaker. Like Hildebrandt and many other contemporaries, he entered architecture after service as a military engineer. He came to Wurzburg as an ensign in the episcopal guard and rose to the rank of cap-

3.62 Sala Terrena of the Belvedere

tain. When Johann Philipp Franz von Schonborn became bishop of Wurzburg, he chose Neumann as his architectural assistant. He soon became the architectural consultant to many of the princes of southern Germany and the designer of a substantial body of executed work.

Schloss Augustusburg in Bruhl has perhaps the most sumptuous of Neumann's stairs, and it was to work on the stair (from 1740 to 1748) that he was employed as a consultant by Clemens August. Neumann's stair is approached from the portico, directly on the axis. The steps seem to fall, like a stream over rocks, from an unroofed grotto at the landing, and to flow into the entrance hall between clusters of white marble caryatids (figs. 3.63 and 3.64). The eye follows the sweep of the stairs upward to left and right toward the oval hole of the artificial sky, diverted for a moment by the monument to Clemens August with figures realistically posed behind a belvedere overlooking the landing. The journey becomes a visual adventure, with fresh architectural incidents presented every moment of the way.

The stair at Bruhl is an imperial stair, the pattern for the grandest of the baroque stairs. Neumann's stair for the Episcopal Palace at Wurzburg (1737–1743) is similar but more dramatic. The design of the Wurzburg palace was an unintended team effort; Hildebrandt prepared plans and so did a group of gentlemen architects under Lothar Franz. The idea for the *cour d'honneur* where Johann Philipp could descend from his carriage at the foot of the great stair seems to have been Hildebrandt's (as were the elevations of the wings), and the overall design seems to have followed the ideas of Lothar Franz's team. Neumann, with the assistance of Johannes Dientzenhofer, was given the day-to-day control of the work, and he seems to have used the opportunity to increase his role gradually; certainly the stair is unquestionably his design.

On the lower floor of the Wurzburg palace (figs. 3.65, 3.66, and 3.67), the staircase walls have been dissolved so the stair floats within arcades on three sides; its long, impressive flight at the bottom appears to rise from within the arcade adjoining the portico. The view is at first constricted and then widens out as one ascends. After an intermediate resting place, the flight continues to the intermediate level landing. From there, two flights return on either side to the vast open space of the piano nobile, which is flooded with light from large windows. Hovering over the whole space is the vaulted ceiling and Tiepolo's fresco. Neumann's son, after his father's death, tells of the rivalry between Neumann and Hildebrandt. In this anecdote, Hildebrandt said he would hang himself from the vault if it remained standing, to which Neumann retorted that he would personally discharge cannons under the structure to prove its soundness (Meyer 1967, 99).

The Wurzburg fresco presents us with its illusory world and sky where Apollo reigns over the kingdom of the arts. The stair emerges on the upper level, surrounded by an ambulatory, which visually expands the volume. The space of the stair is like an inverted pyramid, with the ceiling to carry the vertical space through to the infinity of the painted sky above. Neumann had proposed a frothy rococo decorative treatment for the balustrades, akin to

3.63 Schloss Augustusburg, Bruhl

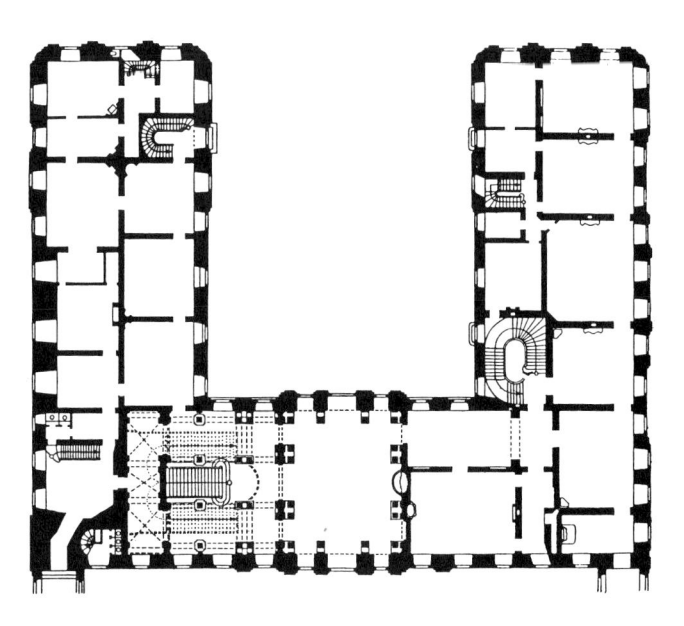

3.64 Plan of the Schloss Augustusburg, reprinted from Mielke (1966), 254

The Baroque Stairs of Honor

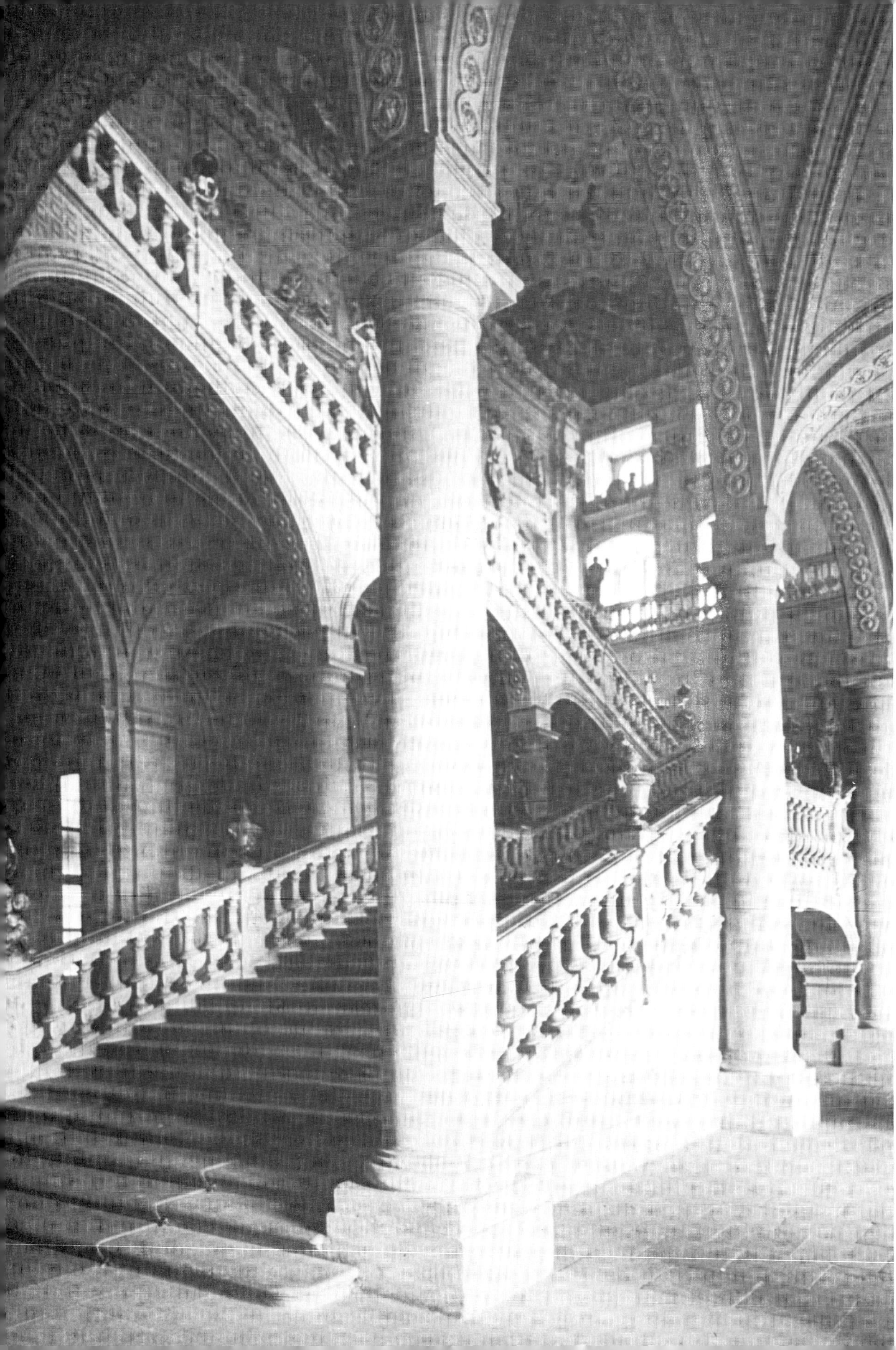

3.65 Episcopal Palace, Wurzburg, entry level

3.66 Plan, Wurzburg palace, reprinted from Mielke (1966), 251

3.67 Imperial stair at Wurzburg palace

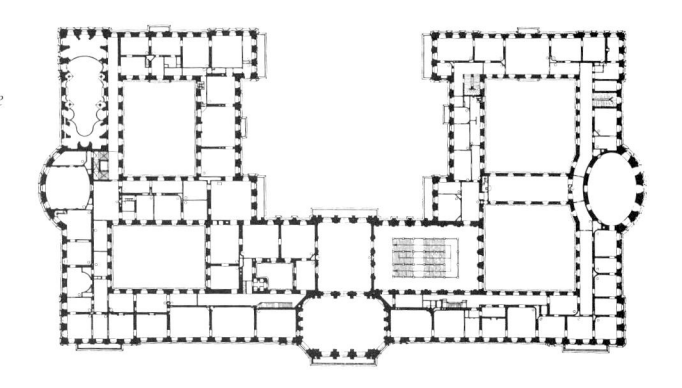

The Baroque Stairs of Honor

Hildebrandt's Mirabell, but abandoned this concept for a more sedate and simpler version.

The architects of these great baroque stairs still struggled to comply with aesthetic goals. The stair as the hub and ceremonial focus of the palace demanded a central position in the plan. Aesthetic doctrine suggested that it should also be set at the center of axes centrally located. The axes should extend in all four directions, with foci carrying the eye out into the gardens and beyond, thus binding the stair to the palace and the garden. The whole composition would demonstrate the total subjugation of nature. In contrast to this ideal, most of these baroque stairs tended to be visual termini enclosed in their stair halls, which limited the flow of space and visual alignment on the main axis. A second problem was that the layout usually prevented one from passing through the stair to the other side of the palace.

One way to address the problem was to start the stair with a matched pair of flights, as at Pommersfelden; one could then see along the axis between the flights, under the landing, and could pass under the landing to other parts of the palace. If the stair doubled back on itself as in an imperial stair, the central returning flight tended to distract the eye from any through view. Schinkel's stair for the Palace of Prince Albrecht in Berlin (1829–1833) is of this pattern and almost succeeds in overcoming the problem through the use of light steel and cast-iron forms (fig. 3.68).

Neumann's proposal for the Hofburg palace in Vienna showed a way out of the conundrum. (The project was never executed, but it became a planning model for the great buildings of the nineteenth century.) His solution was to set the stair in its own wing on the main axis, cut adrift from the corps de logis (fig. 3.69) and in the center of the building complex—a new idea. The wing was to be surrounded by courts, and the space around the stair served to provide a functional connection between the two most important parts of the palace. The stair would have started with two paired flights set at opposite ends of the *Treppenhaus* rather than the single pair of Schinkel. One would have passed between the four flights and under the main landing in order to walk between the two wings of the palace. Alternatively, one could bypass the stair and walk alongside it to the other wing. It would not have been possible to have an overview of the entire stair from the bottom floor; one would have to mount to the main central landing before the whole extraordinary view of the six flights leading up or down would be revealed (fig. 3.70). This would have kept it in conformity with the general baroque model.

STAIRS AND THEATRICAL DESIGN

The influence of theatrical set designers on these baroque architects (and vice versa) is evident. Filippo Juvarra (1678–1736) became the foremost Italian architect of his time after starting his career as a set designer in Rome, and the great nineteenth-century Prussian architect Karl Friedrich Schinkel spent much of his early professional life as a set designer and painter of dioramas. One of the most distinguished set designers from a family of set designers was Giuseppe Bibiena (1696–1756), who spent almost the whole of his working life at the Genoese and Austrian courts. His

Composite Stairs

3.68 Palace of Prince Albrecht, Berlin, by Karl Friedrich Schinkel, reprinted from Mielke (1966), 238

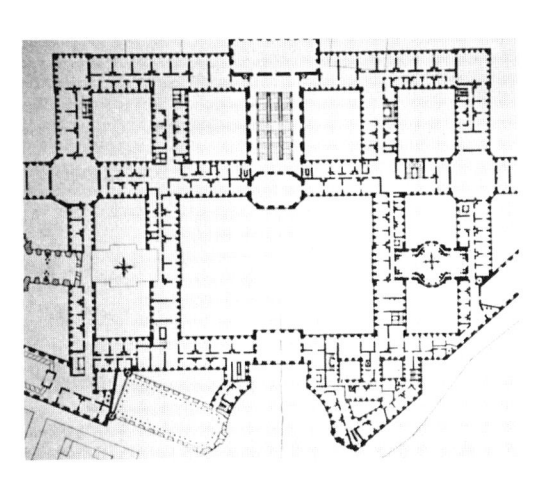

3.69 Neumann's plan for the Hofburg palace, Vienna, reprinted from Mielke (1966), 262

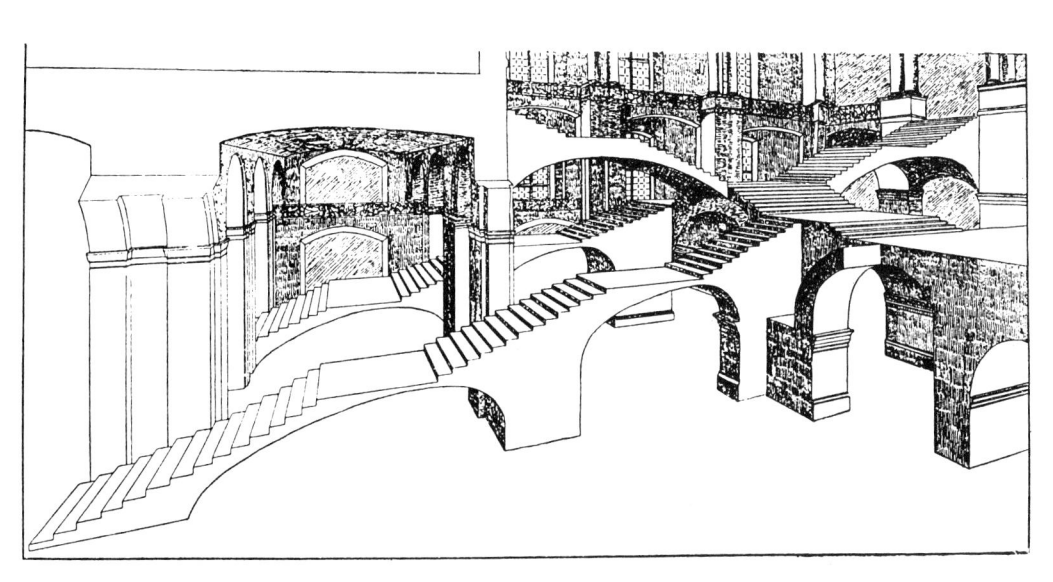

3.70 Neumann's proposal for the Hofburg, reprinted from Mielke (1966), 263

The Baroque Stairs of Honor

perspective designs (fig. 3.71) are full of dramatic stairs passing through elaborate baroque settings into courtyards open to the sky, surrounded by belvederes and viewing galleries.

Sometimes these illusions were even exploited as quadratura. For instance, at the Villa Lechi (1741–1745) in Lombardy by Antonio Turbino, there is an immense architectural trompe l'oeil fresco in the ballroom, depicting a fantastic theatrical stair leading outside and connecting to an arcaded bridging structure at a higher level (fig. 3.72). Like the Bibiena designs, the stair takes the beholder on an architectural journey through a whole compendium of baroque spatial devices.

Straight Flight Compositions

Several less common compositions using straight flight stairs are worth mentioning. The pair of stairs leading up to Palladio's Villa Foscari at Malcontenta (1560) start at right angles to the building (fig. 3.73); after the landing turns through a 90-degree angle, they rise up to the piano nobile against the outer wall of the villa. There is a slightly more complex version of this stair at the external approach to Starke's Palace of the Great Garden (1678–1683) in Dresden. Again the staircase is symmetrical, but this time the first flight of each stair rises from a common central point, runs up parallel to the external walls of the palace, takes a right angle turn at a landing and rises toward the palace, takes another right angle turn at a landing, and rises to meet at a common entrance landing.

Helical Compositions

Curves introduced into the sides of straight stair flights generate the gracious flowing lines that were typical of some of the great

3.71 Bibiena's stair fantasy

3.72 Stair mural, Villa Lechi, Lombardy

Composite Stairs

3.73 Villa Foscari, Malcontenta

3.74 Sans Souci palace, Potsdam

early Renaissance helical stairs, such as Vignola's at Caprarola. Such a device was used, for example, to soften and transform the imperial staircase of Garnier's Paris Opera House; the edges of the lower flights are curved to suggest a spilling out of the treads (and the audience) into the volume of the lobby. A similar device is used in reverse at the top of the stairs. Its ancestor is Michelangelo's Laurentian Library stair. Open air examples date back several centuries. Perhaps the most famous are those of the Sans Souci, Potsdam (fig. 3.74). This pleasure palace, "without a care," as its name implies, was built by Frederick the Great (1745–1748). Here, curving lines are combined with those of the straight flight stairs. From the palace facade, the stair flows down in six flights, each one opening out onto a terrace.

STAIRS WITH HELICAL SEGMENTS

Perhaps the simplest type of stairs composed using helical segments introduces the landing into a geometrically regular helical stair. An example is the oval staircase in the Vitzthum-Schonburg Palace (1774), Dresden, where four landings have been placed on the axes.

Curved stairs and ramps lend themselves even more than straight flight stairs to an expression of the continuous flow of space from one level to another, which parallels the smooth transition of people up or down (fig. 3.75). Symmetrical and opposing curved flights emphasize the movement of the participants on formal occasions with more grace than can be achieved with the use of straight flights—hence, their use externally in the Villa di Papa Giulio and at Villa d'Este. Again it appears that external architecture first exploited these devices.

3.75 Schloss Sleissheim, near Munich

Where the two helical segments are composed with centers on each side of the stair, the product is the serpentine horseshoe stair (Escalier du Fer-à-Cheval) of the Cour du Cheval Blanc (figs. 3.76 and 3.77), built in 1634 to the design of Jean du Cerceau as an addition to the older building of the Palace of Fontainebleau. Guadet (n.d., 329) complains that "the steps are laid out with capricious and dangerous directions." In these stairs, the architect introduced serpentine movement into the staircase, the handrails, and the participants, in classic baroque style. A further example is the stair of Schloss Solitude (1736–1767), Stuttgart (figs. 3.78 and 3.79).

The Baroque Stairs of Honor

3.76 *Cour du Cheval Blanc, Fontainebleau*

3.77 *Plan of the stair of the Cour du Cheval Blanc*

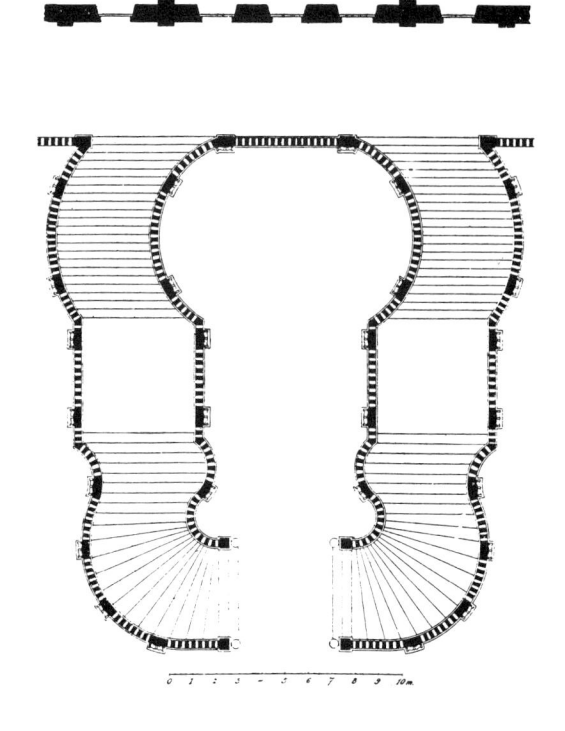

Composite Stairs

3.78 *Schloss Solitude, Stuttgart*

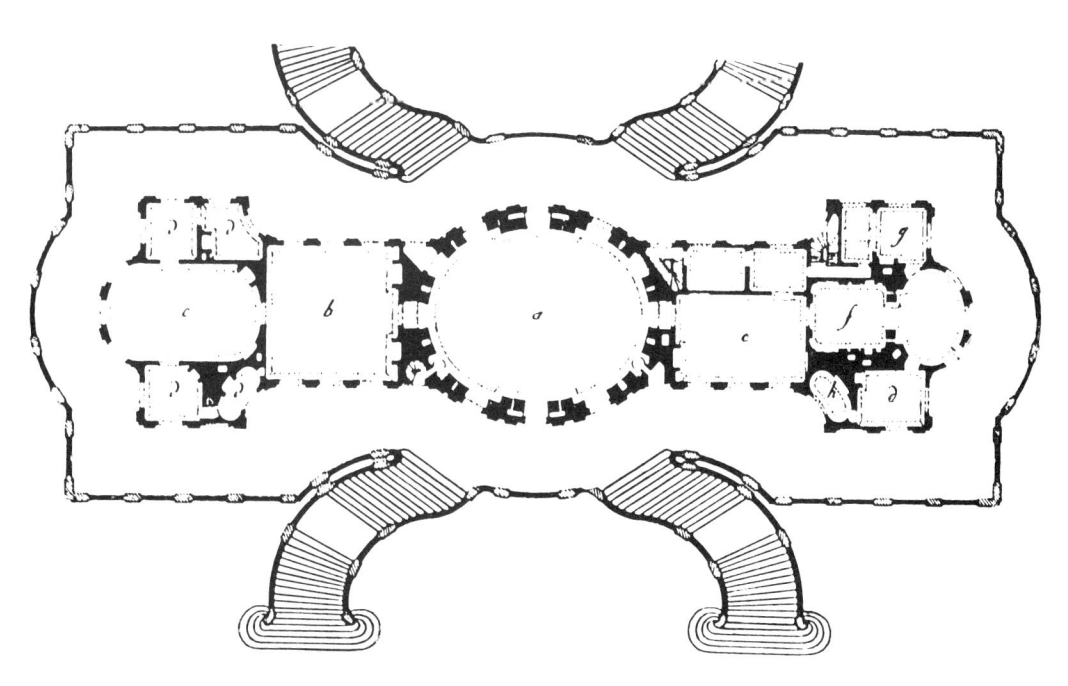

3.79 *Plan of Schloss Solitude, reprinted from Mielke (1966), 139*

THE SPANISH STEPS AND THE RIPETTA

The Spanish Steps in Rome (1670–1723) are an incomparable example of the stair as urban landscape theater. It is the use of a composition of straight and curved segments that forms and enriches this complex sequence of spaces down the Pincian hill (fig. 3.80). The steps take their name from the Spanish embassy, Palazzo Monaldeschi (originally the piazza was called Piazza della Trinità after the church at the top).

It is still uncertain who was the designer.[8] The seed money to construct the scalinata came from a legacy from a French diplomat, Etienne Gueffier, who had been in Rome. Pope Gregory XII had shown an interest in creating a dramatic formal approach to the church since it was completed in 1587. Drawings by Alessandro Specchi exist that are similar to the finished stair (fig. 3.81), and it was he who was the author of that other urban stair fantasy, the steps on the Ripetta (1704), Rome's now-defunct port on the Tiber (fig. 3.82). However, the sketches of Francesco de Sanctis are perhaps the closest to the final product. Girolamo Rossi in 1725 produced an engraving of de Sanctis's design, and this was dedicated to Louis XV, who provided continuing funding for construction. Nevertheless, it was Elpidio Benedetti, Cardinal Mazarin's agent in Rome, who seems to have produced the first sketches (1660), and there is strong evidence that the studio of Gianlorenzo Bernini (1598–1680) may have generated the ideas on which Benedetti's drawing is based.

Regardless of the authorship, the stair is one of the world's most celebrated urban places, so well known that there seems to be little that one can add to describe its character. The stair is designed to be seen

3.80 Spanish Steps, Rome

Composite Stairs

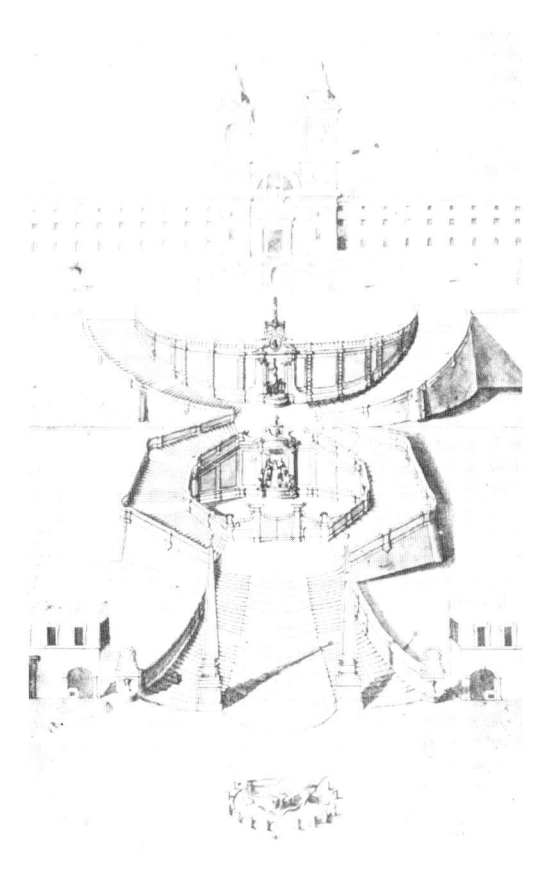

3.81 *Specchi's sketch for the Spanish steps*

3.82 *Ripetta steps, by Alessandro Specchi*

The Baroque Stairs of Honor

from afar as well as from close up and to offer an endless sequence of visual experiences. The landings are as significant as the steps. They are platforms on which to pause to enjoy the play that one enters into by climbing up or down. They are places for the street vendors, fashion models, and photographers, for people to meet, to embrace, to play, and to look. All of these experiences are manipulated within the constantly changing elegant curvilinear design.

Guarini's Palazzo Carignano and Neumann's Bruchsal

Two of the finest baroque stairs are not imperial stairs but helical composites: Guarino Guarini's for the Palazzo Carignano (1679–1692) in Turin and Balthazar Neumann's for the palace at Bruchsal, built some fifty years later. Guarini was the temperamental and talented spiritual successor to Borromini. The palazzo for Emmanuel Philibert, prince of Carignano, is probably the best example of the genre in late-seventeenth-century Italy. The facade, reminiscent of Bernini's original design for the Louvre, is characterized by the sweeping curves that form the protruding central mass, which incorporates the entrance and the staircases.

Inside the entrance is a small foyer that leads into an oval room, which opens onto a courtyard. At either end of the oval room, steps lead to rectangular vestibules. From each of these vestibules, a flight of convex steps follows a curved path, which traces the swelling of the exterior wall, through an enclosing passage to a landing, whose windows provide the only illumination (figs. 3.83 and 3.84). The flights continue after the landing but offer quite a different appearance. The treads now are concave and lead to a brightly lit vestibule opening into a magnificent grand salon. This stair, from an architect who states that "stairs are the most difficult part of a building," is comparable as a spatial tour de force only to that of Bruchsal.

The remarkably beautiful helical composite stair at Bruchsal (figs. 3.85 and 3.86), in the Episcopal Palace (1721–1732), came about almost by chance. Neumann was brought in by Damian Hugo, the prince-bishop of Speyer and the nephew of Lothar Franz (the builder of Pommersfelden), to resolve the architectural tangle that had ensued as a result of his insistence on using architects as consultants but never as supervisors (Laing 1978, 291). Damian Hugo evidently wanted a circular staircase well for the corps de logis of the new palace he was building; perhaps he had Guarini's masterpiece in mind. The design would not have been a problem if he had not belatedly decided to add a mezzanine floor. This mandated an extra half story, with no room for the additional steps.

Before Bruchsal, Neumann had worked at Wurzburg, originally for Johann Philipp Franz—Lothar Franz's other nephew—and later for the then prince-bishop, Christof Franz von Hutten. With Damian Hugo, he redesigned the corps de logis and its core. He managed to fit the stair into the vacant hole. Pevsner (1968, 284) suggests that Bruchsal "with its perfect unity of space and decoration was the high water mark of the Baroque style." Describing the stair as it was before it was damaged in World War II and restored after the war, he says:

The arms [of the stair] started in the rectangular vestibule. After about ten steps one enters the oval. On the ground floor it is a sombre room

3.83 *Palazzo Carignano, Turin*

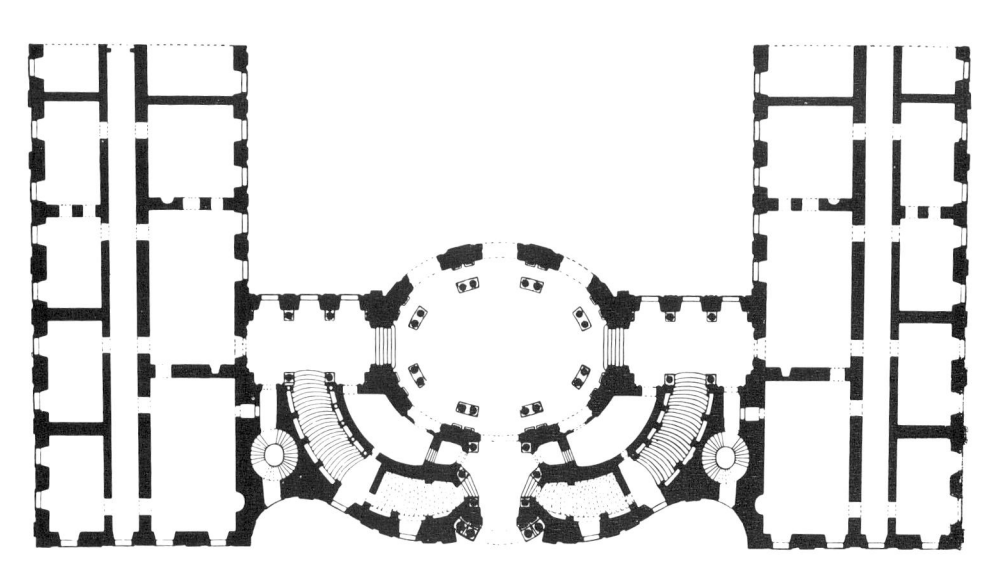

3.84 *Plan of Palazzo Carignano*

The Baroque Stairs of Honor

152

3.85 Bruchsal stair

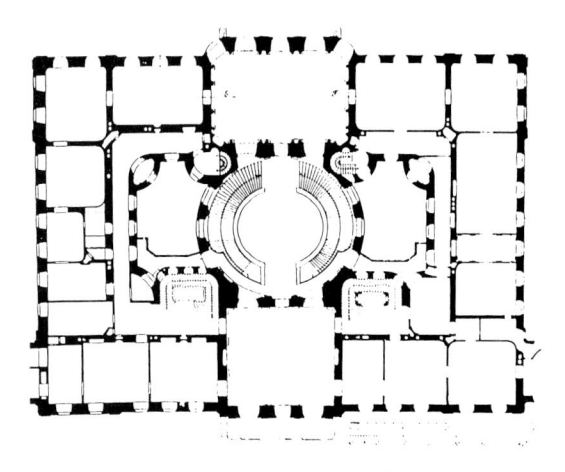

3.86 Bruchsal plan, reprinted from Mielke (1966), 85

Composite Stairs

painted with rocks in the rustic manner of Italian grotto imitations. The staircase itself then unfolded between two curved walls, the outside wall solid, that on the inside opened in arcades through which one looked down into the semi-darkness of the oval grotto. The height of the arcade openings of course diminished as the staircase ascended. And while you walked up, it grew lighter and lighter around you, until you reached the main floor and a platform the size of the oval room beneath. But the vault above covered the larger oval formed by the outer walls of the staircase. Thus the platform with its balustrade separating it from the two staircase arms seemed to rise in mid-air connected only by bridges with the two principal saloons. And the vast vault above was lit by many windows, painted with the gayest frescoes and decorated with a splendid fireworks of stucco. The spatial rapture of the staircase was in this decoration transformed into ornamental rapture.

FERDINANDO SANFELICE

The Neapolitan architect Ferdinando Sanfelice (1675–1748), another of Balthazar Neumann's contemporaries, was also a master of stair design. His clients were not the noble princes of the church and the state, so his palaces are far more modest in size and decoration. Naples, moreover, was not the center of a great kingdom. It was simply a province of Spain that was ceded to Austria in 1713, when Sanfelice was about thirty-eight, as a result of the War of Spanish Succession. This political event brought Austrian influence, ideas, and capital. Naples was a crowded city, so the palaces built by Sanfelice and his colleagues, like those in Genoa, were urban edifices. These were frequently built on relatively small sites and were up to five stories high.

For the palaces that he designed, Sanfelice exploited the open stair that was popular in the area and climatically suitable. He gave the staircases of his urban palaces a new grandeur and new forms that still are baffling in their formal and spatial manipulation and complexity. He set his stairs in the place of honor as the dominant feature occupying the whole end of a courtyard, no longer tucked away in a corner, and he accepted and delighted in the integration of the ensuing diagonal lines. The Palazzo Sanfelice (1728), designed for his own family, was one of the largest of these palaces and included two courtyards. Forming the end wall of the right-hand courtyard was a sort of multistory imperial staircase that climbs up the whole facade (fig. 3.87), almost a precursor of the effect of nineteenth- and twentieth-century fire stairs. The stair is expressed on the outside face in a way reminiscent of Jakob Prandtauer's stair facade (fig. 3.88) for the Benedictine monastery of Saint Florian (1706–1714) near Linz, Austria. By 1728 Naples was ruled by Austrian viceroys, so it is quite likely that the architects were familiar with each other's work.

At the Palazzo Sanfelice, one enters the stair and turns left or right, parallel to the facade, and ascends to a landing. Flights return parallel to the first-floor flights, ending at a landing at the central bay. A final flight, in the direction of the entrance and at right angles to the first two, leads on to a place immediately over the entry point. Because the central bay is open, it offers a view between the courtyard and a garden or grotto on the second floor.

There is an equally ingenious stair in the other courtyard of Palazzo Sanfelice, which has features in common with another Austrian stair, the double helical stair at Graz (fig. 2.40). Sanfelice's version consists of

3.87 Palazzo Sanfelice, Naples

3.88 Benedictine monastery of Saint Florian, near Linz

Composite Stairs

two helical stairs that touch each other (fig. 3.89) and share common landings at the first and second levels. One enters by a short straight flight stair leading to a trapezoidal landing. From there, one elects to enter a helical flight on the right or left; the flights are about 2 m wide. After ascending one quadrant of either flight, one is at a triangular landing off which a door leads to the mezzanine. Continuing up the chosen flight, one finds oneself at a landing shared by the two adjoining stairs. From there, one climbs around another hemicircle to the doors of the piano nobile.

Several other Neapolitan palace stairs of Sanfelice invite attention (and there are many of his stairs that merit inclusion): the Palazzi Serra di Cassano, Bartolomeo di Maio, and Capuano. The stair of the Palazzo Serra di Cassano is in many ways Sanfelice's most spectacular. It sits within a great stair hall open to the approach courtyard. After entering through the massive arch, one passes between ascending flights on both sides (fig. 3.90) to a landing immediately below the main entrance to the piano nobile. From this landing, one can choose to go left or right up several steps to a landing, where flights ascend toward the entry courtyard and end in semicircular landings from which steps dogleg back in the opposite direction. From the landings at the top of these flights, the final flights lead off at right angles, ending at the top landing at the main entrance. The stair is

3.89 Helical stair at Palazzo Sanfelice

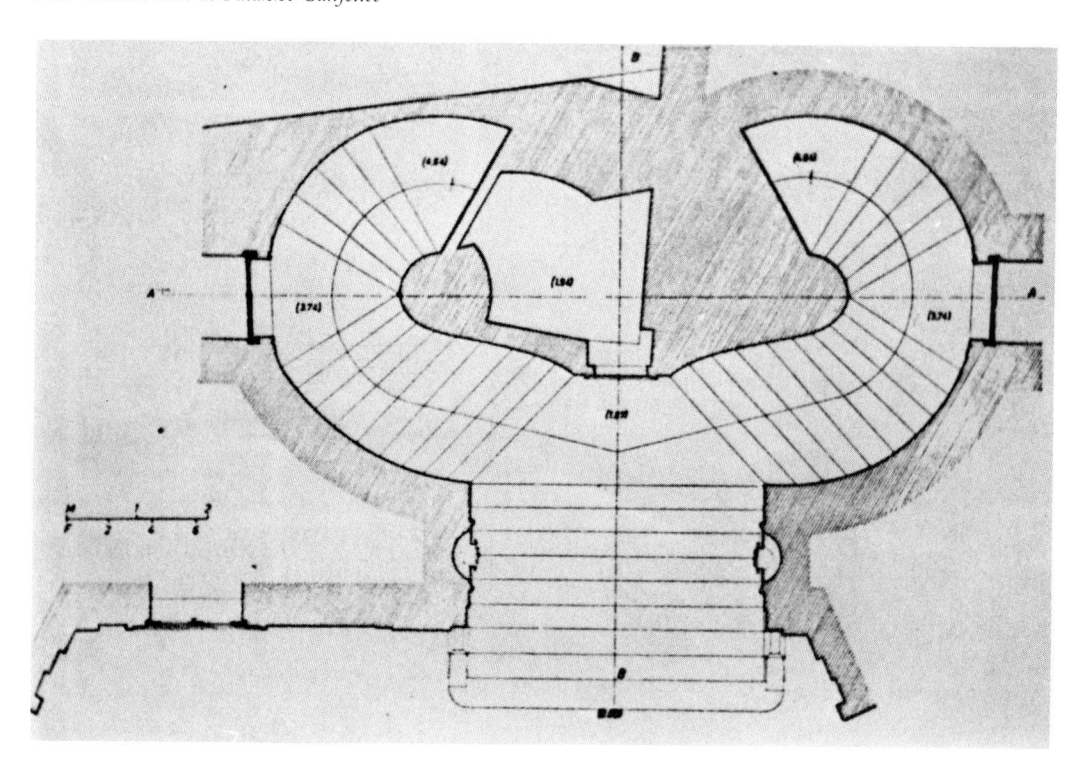

3.90 *Palazzo Serra di Cassano, Naples, the octagonal stair*

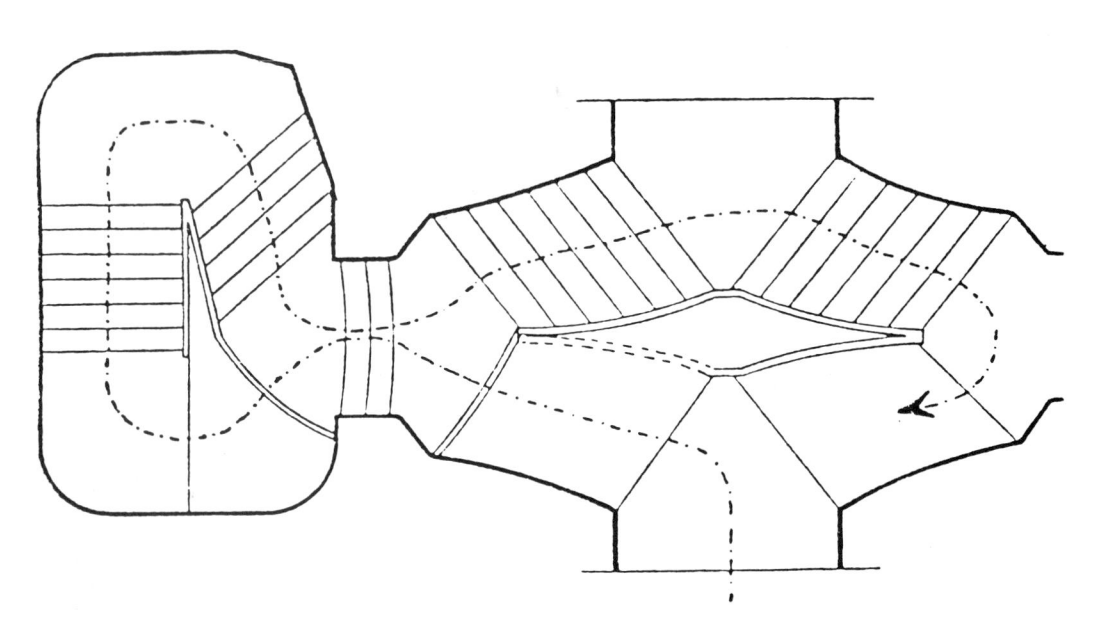

3.91 *Palazzo Bartolomeo di Maio, Naples, octagonal windows*

Composite Stairs

beautifully constructed from a dark gray stone with balusters and other details in a cream-colored marble.

The stair for the Palazzo Bartolomeo di Maio on Discesa della Sanità is composed of two linked open well helicals (fig. 3.91) where the flights curve inward toward the well. The space available for the stair is quite small; this unusual stair ingeniously crams in a complicated journey during the ascent.

At the Palazzo Capuano, Sanfelice designed a compact octagonal staircase set at the end of a rather narrow court (fig. 3.92). The octagon projects into the court without blocking adjoining windows. It is a cousin of the main stair in his own palace but rather more compact. After entering from the courtyard, one may set off at 45 degrees to left or right up a flight leading to a triangular landing. A 45-degree turn leads to a short flight leading to an entrance; a 90-degree turn leads to a flight up to a shared landing from which one must turn through 135 degrees and up a flight to a landing over the place where one entered the stair. The octagonal plan shape is repeated in the window openings (fig. 3.93), thereby concealing the diagonal ascent line of the stairs from the facade.

Not far from Sanfelice's Naples lies the immense Carthusian monastery of Padula. The building is in the form of a gridiron, symbolizing the martyrdom of St. Lawrence after whom it is named. The grand staircase, built in the second half of the seventeenth century to the design of Gaetano Barba, represents the bowl of grease

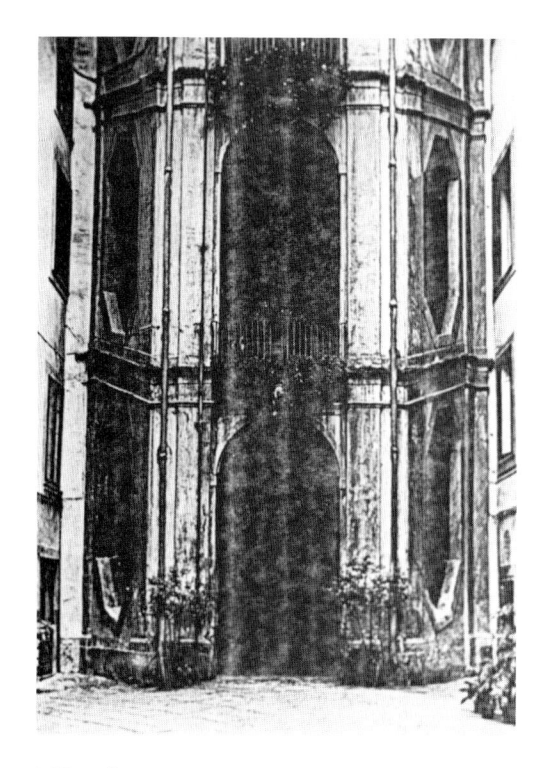

3.92 Palazzo Capuano, Naples

3.93 Palazzo Capuano

The Baroque Stairs of Honor

of the martyrdom. The stair has two opposing flights, which sweep up to landings before returning and rising to the gallery above (fig. 3.94). Like Sanfelice's stairs, large openings illuminate the flights and connect them visually with the outside (fig. 3.95).

In the late eighteenth century and early nineteenth century, celebration of the flowing curved lines of such staircases took two forms: an emphasis on flow and counterflow in such complex designs as the stairs of the Wurzach Palace (1723–1728, by an unknown architect) in Germany (fig. 3.96) and a concentration on the achievement of greater simplicity in England's Georgian and Regency stairs, in which a single flight links two levels. Some of the most graceful of these composite helical stairs have quite modest dimensions, particularly in Georgian England in the elegant town houses. For example, in the Royal Crescent at Bath by John Wood the younger (1767–1775), the stair of a typical house is obvious but not obtrusive and small but quite adequate for the house (fig. 3.97). It sweeps up helically in a 90-degree turn, rises in a straight flight, and finally sweeps into another 90-degree helical turn at the next floor.

Long after reception rooms had ceased to be located on the second floor, the grand stair and hall remained popular, for by this time the stair had become part of the trappings of status. There are numerous examples of preposterous and ostentatious stairs in quite small villas.

3.94 *The Charterhouse, Padula*

3.95 The Charterhouse

3.96 Wurzach Palace, Germany

The Baroque Stairs of Honor

3.97 House in the Royal Crescent, Bath

3.8 The Nineteenth and Twentieth Centuries

As the Age of Enlightenment drew to an end, extravagant palaces and their stairs lost much of their administrative relevance and status; in France the baroque became a despised symbol of the excesses of the monarchy. After the American and French revolutions, the new republics espoused the democratic ideals of the golden age of Greece. A renewed, more restrained classicism seemed appropriate for the important new buildings of the new republics and the increasingly democratic monarchies. The power and worthiness of the modern state was now demonstrated by the instruments of the state rather than those of absolute monarchs. The new palaces were the public buildings—museums, universities, courthouses, opera houses, and theaters.

Straight flight stairs became fashionable again. They were made to be appropriately grand without extravagance, emphasizing now the eminence and power of the state. Their generous dimensions, no longer required for ceremonial occasions, were necessary for the burgeoning crowds. Civic buildings and governmental palaces were constructed for the express use of the proletariat that now thronged into these public realms. Crowd management became one of the necessary design considerations. Spatially, these public stairs, bereft of their princely pomp and splendor, were often strictly contained and constrained within enclosing walls that blocked the view and limited the spatial architectonics. Friedrich von Gaertner's design for the State Library (1832–1843) for Munich (fig. 3.98) is fairly typical of the genre. The stair of the library is centrally located in its own wing, following Neumann's pattern for the Hofburg

Composite Stairs

3.98 Munich State Library, reprinted from Mielke (1966), 272

Palace, to provide convenient access to all parts of the building.

Unlike the baroque palaces, these state and municipal monuments were designed to facilitate entrance to much of the building, not to restrict admission to a few reception rooms on the piano nobile. Therefore, the main stairs gave access to all the public floors. A. von Wieleman's Palace of Justice in Vienna (1875–1881) is a fine example (fig. 3.99). It swoops up from the glass-roofed atrium, leading the eye (and the ascending visitors) toward a 10-foot-high marble statue of justice. Stairs in these monumental buildings became dramatically higher, serving many floors. For example, the Munich Palace of Justice by Friedrich von Thiersch (1890–1897) has grand stairs rising three floors on two sides of a glass-roofed atrium.

The technical advances in the use of cast iron, steel, and glass made it possible to explore once more the possibilities of the artificial sky. The skylight allowed unbroken walls, free from windows. It was no longer necessary to paint the sky; the increasingly large atriums could be roofed with naturally illuminated glass domes. Michelangelo's wonderful conception of a skylit courtyard as the vestibule for the Laurentian Library became a delightful model for nineteenth- and twentieth-century buildings. These atriums provided all-weather gathering and meeting places. Today the grand stair has been replaced by escalators and glass elevators and the scale of the atriums has soared into the space of an internal skyscraper.

The Nineteenth and Twentieth Centuries

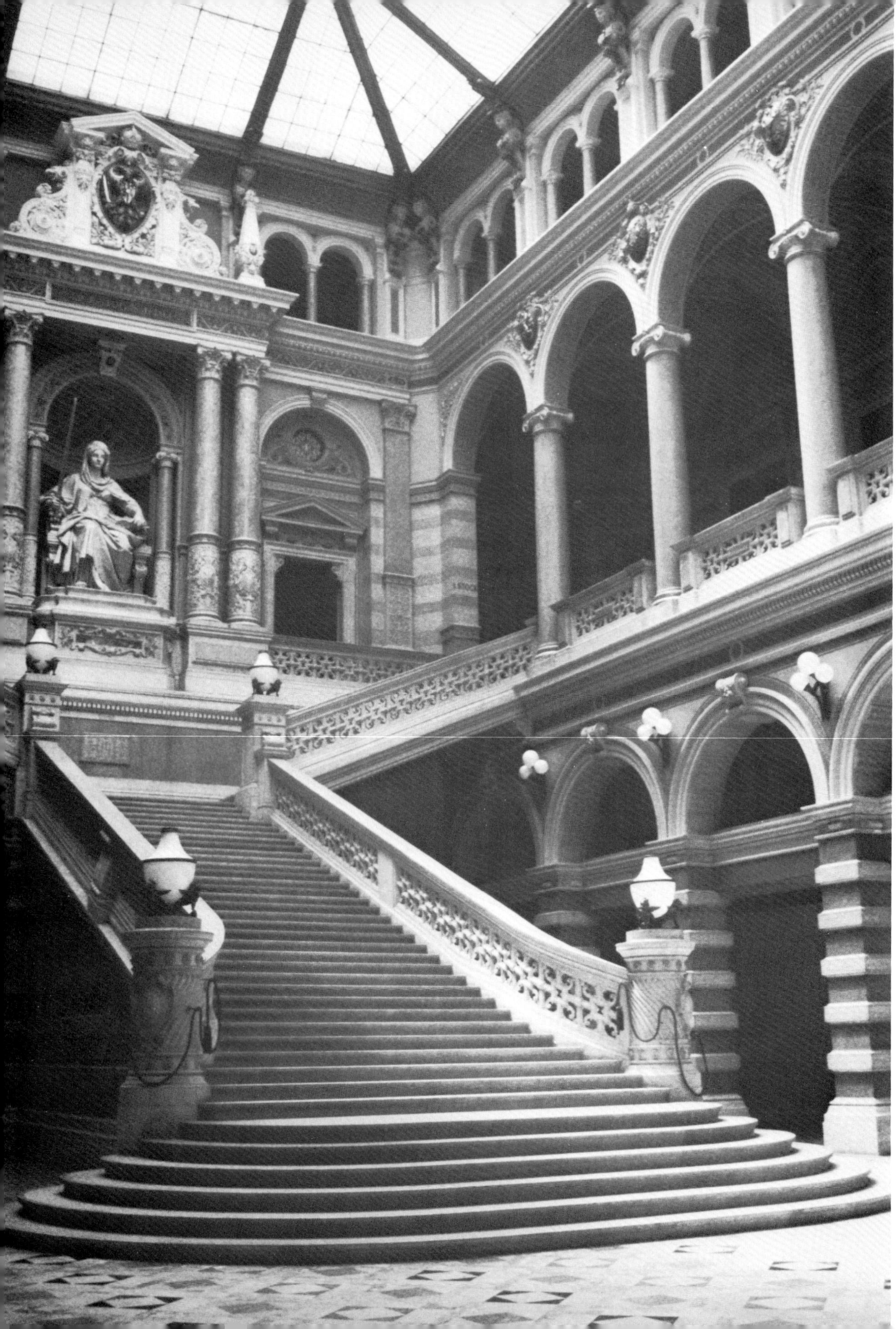

Stairs for Great Crowds

With the increasing participation of the masses in the affairs of government and state, public buildings were expected to welcome large crowds of people routinely. The staircase as part of the mass movement system had to move the crowds quickly and safely to the accessible parts of the building. Stair capacity became as important as location and effect. Schinkel's Altes Museum in Berlin (1824–1828) featured the stairs, which are the central component of the movement system (fig. 3.100). The stairs were made into part of the main facade, appearing on the second level.

Perhaps the most influential effect on stair design grew from reactions to several disastrous fires in the nineteenth century. Mielke (1966, 289) lists, among others, the 1836 fire at the Lehmann Theatre in Petersburg (800 died), the 1845 fire at Quebec's Théâtre Royal (500 died), the 1846 fire at Canton Theatre in China (1400 died), and the 1876 fire at Brooklyn's Convoy Theatre (283 died). As a reaction to these and other public safety concerns (such as a better understanding of the transmittal of infection), comprehensive public health laws and building and fire codes to mandate minimum standards of safety were enacted in many countries.

Gottfried Semper (1803–1879), the most significant German architect of the second half of the nineteenth century, demonstrated this new concern for safe fire evacuation in his design for the Vienna City Theater (1888). This is both a functional and safe solution, with no fewer than seven enclosed stairs separating the loggia from the foyer and another one at each end of the loggia and the foyer (fig. 3.101).

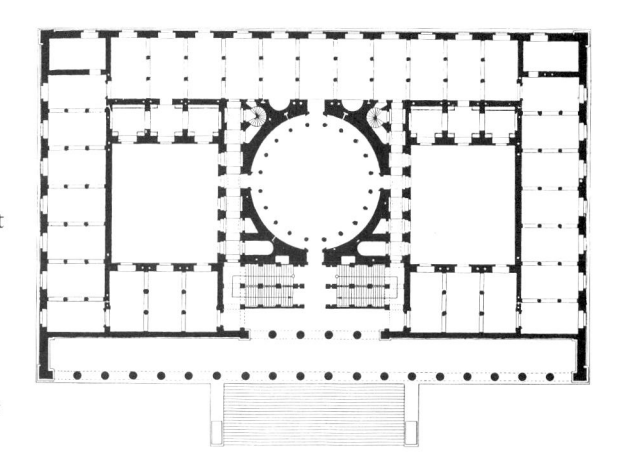

3.100 *Altes Museum, Berlin, reprinted from Mielke (1966), 285*

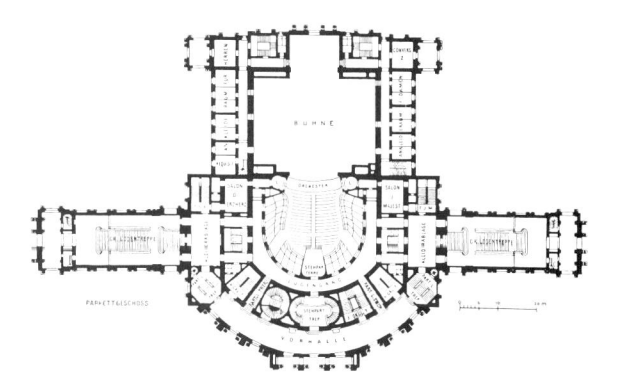

3.101 *Vienna City Theater, reprinted from Mielke (1966), 290*

3.99 *Palace of Justice, Vienna*

Jean-Louis-Charles Garnier (1825–1898) set the pattern for resolving the conditions required by the codes while introducing a radically different way of thinking about theater foyers. Garnier was only thirty-five years old when he won the competition for the design of the new Paris opera house; Viollet-le-Duc was one of the also-rans. Garnier's proposal was so unusual that it was strenuously attacked by the critics, including Viollet-le-Duc who censured Garnier for not subordinating the stair hall to the main auditorium.[9] This was, of course, the brilliance of Garnier's design (figs. 3.102 and 3.103). He realized that it was possible to intensify the architectonic grandeur of the foyer and still serve the functional needs of a great crowd of opera patrons. This might seem to us to be an obvious solution, but it was a major contribution to nineteenth-century theater design and became the model. His staircase is the architectural as well as spatial center of the building. He recognized that theater patrons enjoy the social scene of the foyer as well as the performance on the stage. Unlike Semper's theater stairs (and the great baroque stairs), Garnier's grand stair gave access to all the public parts of the building. Concurrently they provided the users and the watchers a stylish spectacle set against neobaroque splendor and sensory manipulations. "At every floor," Garnier wrote in *Théâtre National de l'Opéra de Paris*, "the spectators leaning against the balconies line the walls and bring them to life, while others going up and down add still more to the scene. . . . The sparkling lights, the resplendent dress, the lively and smiling faces, the greetings exchanged; everything will contribute to a festive air, and everyone will enjoy it without realizing how much the architecture is responsible for this magical effect."

Victor Hugo loved the building, calling it "comparable to Notre Dame." Claude Debussy loathed it: "Everybody knows the Paris Opéra . . . at least by repute. To my regret, I have been able to assure myself that it has not changed; to the ignorant passerby it still looks like a railway station; inside it is very much like a Turkish bath. Odd sounds, called music by those who have paid to hear, are still produced there . . . its going up in flames would be no bad thing but for the harm the innocent people would undeservedly suffer."

Garnier knew that his Opéra was a Parisian landmark. He declared, "The Opéra is the staircase just as Invalides is the dome and Saint Etienne-du-Mont is the rood-screen."

THE TWENTIETH CENTURY, MODERNISM, AND ON
The development of the elevator and the escalator in the nineteenth century had a profound effect on architecture. They made the skyscraper practical; or, perhaps more correctly, the maturation of structural science, advancements in steel and reinforced concrete, and the economic pressures in big cities mandated the invention of mechanical vertical movement systems to exploit the construction possibilities of these materials. The escalator and the elevator paralleled the invention of quite new building types—the high rise and the mass movement building. They did not, however, replace the stair. They added two new movement systems with functions that are quite different from those of any stair; each has its own functional and economic zone of efficiency.

Composite Stairs

3.102 Paris Opéra

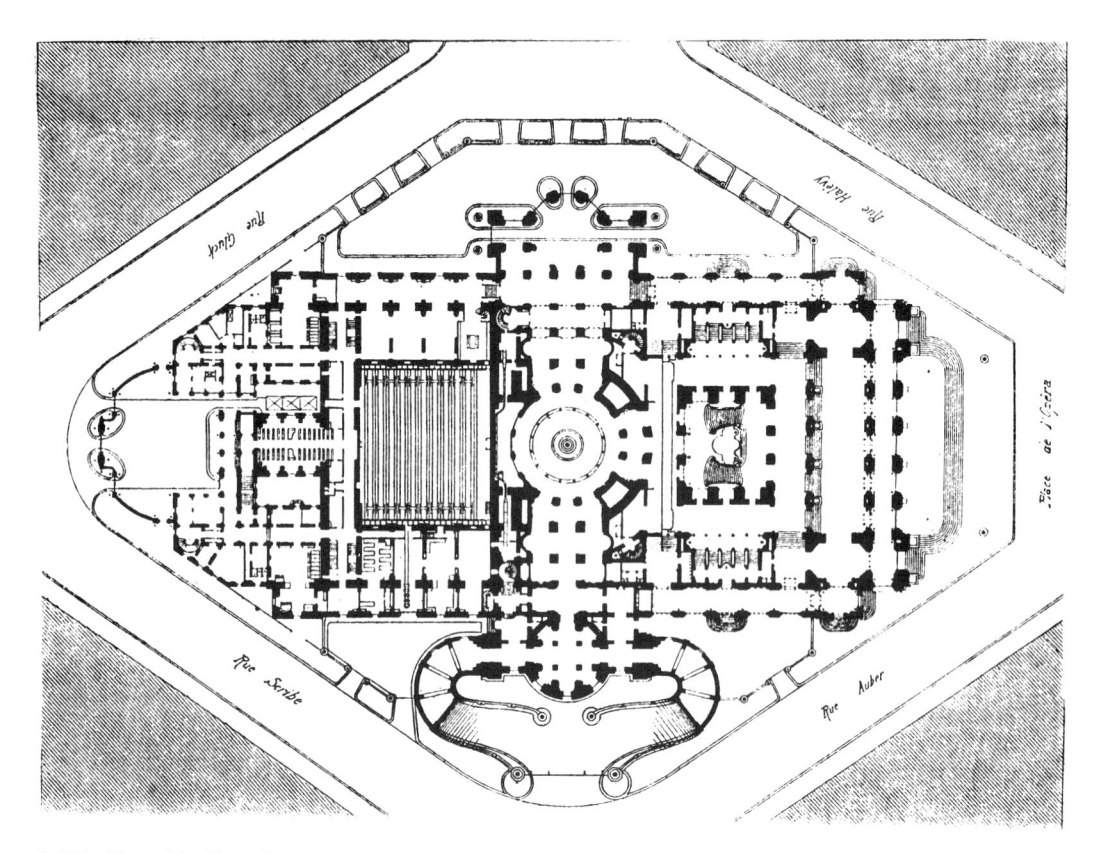

3.103 Plan of the Paris Opéra

Elevators were developed for and remain the only feasible primary movement system for high-rise buildings. The stair has been used for providing access and egress for as many as six floors (as in Paris), but today three floors is probably a more usual upper limit. The stair's ability to move large crowds safely and quickly is not great. Ramps are safer, and escalators are much more effective for moving crowds swiftly up or down, particularly when the distance to be traversed is more than one floor. In the vast twentieth-century atriums and even department stores, the escalator and the elevator have become the dominant architectural symbol of movement in a faster and more crowded world. This is as much related to the scale of these huge spaces as to the efficacy of the mechanical transportation; the glass elevator and its vertical shaft, and the powerful diagonal line of long escalators (fig. 3.104), are able to contribute as compositional components of very large spaces in a way that is not possible for the stair. The stair's scale giving ability also limits it to somewhat smaller spaces.

Modernism may be thought of as synonymous with functionalism (which is to confuse one of the slogans with the product). There was an emphasis on the practical, but this was never intended to excuse the uninteresting and the nonarchitectural. As a rule, however, the stair tended to be subordinated within spatial and architectonic compositions; although seldom the dominant element, it was often used to give vitality, accent, and interest within the canons of abstraction.

Certain characteristics are often typical of the stairs of modernism and reflect the social, technological, theoretical, and architectonic attitudes and goals. Perhaps more than in any other period, the nineteenth- and twentieth-century stair has been explored as a spatial-structural-aesthetic object. Cheap cast iron opened the way to delicate as well as decorative and decorated structures. Steel, concrete, glass, and plastics made it possible to extend the goal of dematerialization that had begun a thousand years earlier. For modernism, this dematerialization was part of the greater philosophical goal of extending and enlarging the apparent space of the room by the use of transparency—to make the space appear larger by reducing solid parts to a minimum. Arne Jacobsen's graceful dogleg stair for the town hall at Rodovre, Denmark (fig. 3.105), for example, is clearly intended to be transparent, if not invisible. It seems to hang delicately in space and to occupy far less visible space than it does. The observer is dazzled by the structural elegance and economy of materials and shape; less is (was) more.

For this stair and many other twentieth-century examples, structural shock as a response to structural pyrotechnics was often in part an end in itself. The beholder was, and was intended to be, astonished that the structure stood. The helical stair, for example, is by nature akin to the spring in shape; Franco Albini's polygonal helicoidal stair for Palazzo Rosso in Genoa (fig. 3.106) exploits this image. The stair is like a complex of metal springs composed into what appears to be a mobile caught at a moment of stasis—a structural wonder.

The modernist goals of dematerialization and structural excitement were often used to draw attention to the stair in the compositions. Because of its helical or diagonal

3.104 Pompidou Center, Paris

3.105 Rodovre town hall, Denmark

Composite Stairs

3.106 Palazzo Rosso, Genoa

direction, the stair tends to draw attention to itself if it is manifest. In fact, this characteristic was often exploited to give interest to an otherwise static composition. Within the modernist goal of manifest truth, stairs were often revealed and therefore compositionally significant (the exception, as always, was service and fire stairs). To fulfill this role, the stair had to be interesting and sculptural. Because replication was viewed as the sin of plagiarism and tradition was eschewed, the stair had to be visually "new" in some way.

Stylistically, modernism was fascinated with industrial and maritime structures. Balustrades were copied from ship companionways like that of the historic steamship *Eureka* moored in San Francisco (fig. 3.107). The bare-bones simplicity of industrial access stairs was more powerful as an image than the heavy decorated classicism of the late nineteenth century. This showed the way for expressing fire stairs, for example, instead of hiding them in the building carcass.

The undoubted importance of the elevator and escalator in the late nineteenth and the twentieth centuries has tended to overshadow the architectural significance of the staircase. However, the stair continues to be an important element and to be representative of the cultural, architectonic, and functional purposes that it must serve, many of which are not new. Nevertheless, the Zeitgeist has not abandoned us. The stair within the vision of the major architectural movements of this century has continued to evince characteristics that are new and distinctive.

3.107 Companionway of the steamship Eureka

NOTES

2 The Helical Stair

1. Mary Whiteley (1985, 16) suggests that, contrary to the speculations of Viollet-le-Duc, the stair was square rather than circular, with a solid rather than hollow newel, and without the second stairway that he thought was built within the hollow newel.

2. Lise Bek (1985, 118) argues that Alberti's aversion is a polemical criticism of courtly tradition in that he rejects the stair's (secondary) function as a link between areas of differing importance in the building. She suggests that as a humanist, he opposes an audience ceremonial that limits the doctrine of equality—between host and guest, for example—and advocates (architectural) equality "for a harmonious interrelation between free men." This might be a more persuasive argument if Alberti was a product of the baroque. In book V, Alberti himself says of the house, "the principal Parts may be allotted to the principal occasions; and the most honorable, to the most honorable."

3 Composite Stairs

1. Wilkinson (1975) details the three stages of development as the first flight was increased from a single bay to a double bay and ultimately to its final three-bay form, under the personal interest of Prince Philip. Covarrubias's stair was designed to occupy five bays rather than the much grander nine bays that were finally built.

2. Wilkinson (1975) discusses these events in considerable detail.

3. Evidence produced in lawsuits of the time "charged that the Cardinal's [Ippolito d'Este's] archaeologist Ligorio showed no respect for private property, that he had owners imprisoned or banished." David R. Coffin, *The Villa d'Este at Tivoli* (Princeton: Princeton University Press, 1960), 8.

4. Sohm (1985) mentions comparable political activities on stairs of similar form at the Palazzo dei Consoli at Gubbio, the Palazzo dei Priori at Perugia, and the Palazzo Comunale at Cortona.

5. According to McAndrew (1980, 99) Rizzo was a scoundrel who managed to steal some 12,000 of the 80,000 ducats budgeted for construction at the palace.

6. As Wilkinson (1975) has pointed out, the design of this palace has been traditionally attributed to Rocco Lurago. Recent studies suggest that the building may have been designed by Domenico and Giovanni Ponsello.

7. These are but the most important examples in an age of great staircases. One cannot dismiss the particularly rich heritage of northern Italy and Sicily, particularly the villas of Piacenza, Florence, Palermo, and Bologna and the works of Torreggiani, Piacentini, Ruggieri, Silvani, Giganti, and Arrighi (discussed by Wittkower 1968, 390–392, 400).

8. For a fuller discussion on this topic, see Marder (1962) and Gillies (1972).

9. For a fuller discussion on this dispute, see Kodre (1983).

Glossary

Balanced step (or dancing step): a winder built so that its narrow end is only a little narrower than the wider end. It is therefore more comfortable to walk on than a winder in which the nosings radiate from a common center.

Baluster: a post in a balustrade of a flight of stairs that supports a handrail (from the Italian *balausto,* the wild pomegranate whose double-curving calyx tube the baluster often resembled).

Balustrade: the whole infilling from handrail down to floor level at the edge of a stair.

Banister: a baluster (corruption of *baluster*).

Bracket baluster: a (steel) baluster bent to a right angle at its foot and built into the side of stone or concrete stairs.

Bracketed stairs: stairs carried on an open string. The overhanging nosings are usually ornamental.

Carriage (or carriage piece, rough string, bearer, stair horse): an inclined timber placed between the two strings against the underside of wide stairs to support them in the middle.

Cat ladder (or duck board): a ladder or board with cleats nailed on it, laid over a roof slope to protect it and give access for workers to repair the roof.

Circular stair: a helical stair.

Closed stair (or box stair): a stair walled in on each side and closed by a door at one end.

Close string (or closed string): a string that extends above the edges of the risers and treads, covering them on the outside.

Commode step: a riser curved in plan, generally at the foot of a stair.

Corded way (or *cordonata*): a path on a steep slope, protected from erosion by steps formed with wooden or stone risers.

Corkscrew stair: a spiral stair.

Cuner: the wedge-shaped sections into which seats are divided by radiating passages in ancient theaters.

Curb stringer (or curb string): a three-member outer string, consisting of a close string carrying the stair, surmounted by a molding (called the shoe rail) from which the balusters rise, and faced by a facing string.

Curtail step (or round or scroll step): a step curved in plan, so that one or both ends project in a semicircular or spiral shape, usually used for the lowest steps in a flight.

Dancing step: *see* Balanced step.

Dextral stair: a stair that turns to the right during ascent.

Dog-legged stair (or dogleg): a stair with two flights separated by a half-landing, and having no stairwell, so that the upper flight returns parallel to the lower flight.

Double-return (or side flight) stair: a stair with one wide flight up from the lower floor to the landing and two flights from the landing to the next floor.

Escalade: the act of scaling a rampart, by means of ladders.

Flier (or flyer): a rectangular tread.

Flight: a series of steps between landings.

Geometrical stair: (1) a stair with a string that is continuous around a semicircular or elliptical well, and thus has no newel posts and often no landings between floors; (2) "the term commonly applied to stone staircases radiating on their plan from one or more centers, with an open well; the broadest end of each step being tailed into the wall and resting on the rest below it by a back rebate, as in straight staircases, and having in addition to these two elements of strength a third, viz. that derived from the keying character of each step with reference to the adjacent one above and below it." (Architectural Publication Society)

Going: the horizontal distance between two successive nosings. (In a helical stair the going varies.) The sum of the goings of a straight flight stair is the going of the flight.

Gradient of a stair: the ratio between going and riser; the angle of inclination.

Guardrail: a protective railing designed to prevent people or objects from falling into open well, stairwell, or similar space.

Half-space landing (or half space or half landing): a rectangular landing of width equal to two flights (*see also* Quarter-space landing).

Handrail: a rail forming the top of a balustrade.

Handrail bolt (or joint bolt): a bolt threaded at both ends. A square nut at one end is gripped in a mortise in an end of one handrail. In the other handrail, a similar mortise is provided, but the nut is circular and notched and can be turned by striking the notches with a handrail punch inserted into the mortise from beneath the handrail. The tightening of the nut brings the two ends of the handrail close together.

Handrail scroll: a spiral ending to a handrail.

Helical stair: the correct but not the usual name for a spiral stair.

Hipping of handrail: a characteristic of early Georgian wooden balustrades. The handrail rises at the intermediate newel either at a right angle or in a sweeping curve.

Kilt: *see* Wash.

Kite winder: the central of three winders turning a right angle.

Landing: a platform at the top, bottom, or between flights of a staircase.

Margin: the distance between the nosing and the top of an upstanding string.

Monkey tail: a downward scroll at the end of a handrail.

Mopstick handrail (or mopstick): a handrail that is circular except for a small flat surface underneath.

Newel (or newel post): the post around which wind the steps of a circular stair. Also applied to the post into which the handrail is framed.

Newel cap: a wooden top to a newel post.

Newel joint: a joint between newel and string or handrail.

Nosing: the front and usually rounded edge to a stair tread. It frequently projects over the riser below it.

Nosing line: a line touching the lead edges of the nosings of successive treads of a stair. The margin of a close string is measured from it.

Nosing overhang: the distance that the nosing edge of a step projects beyond the back of the tread below.

Glossary

Open newel stair: a geometrical stair (one without newels).

Open stair: a stair that is open on one or both sides.

Open string (or cut string): a string that leaves the ends of the treads and risers exposed on the outside.

Open well stair: a stair with two or more flights around an open space.

Perron: a dignified exterior stairway, usually approaching the main entrance.

Piano nobile: the principal floor of a house, raised one floor above ground level.

Quarter-space landing (or quarter space): a platform of width equal to one flight, where a 90-degree turn is made.

Radial step: a winder.

Ramp: an inclined plane for passage of traffic.

Riser: the upright face of a step.

Riser height (or rise): the vertical distance from the top of a step at the nosing to the top of an adjoining step at the nosing.

Sinistral stair: a stair that turns to the left in ascent.

Solid newel stair: a spiral stair of stone in which the inner end of each step is shaped to form a nearly continuous cylinder with the inner ends of the other steps.

Spiral stair (or helical stair): a circular stair in which all the treads are winders.

Stair: (1) a series of steps with or without landings, giving access from level to level;
(2) one step, consisting of a tread and a riser.

Stairhead: the top of a stair.

Stairway: a staircase, or a stairwell.

Stairwell: *see* Well.

Step: one unit of a stair, consisting of a riser and a tread. It may be a flier or a winder.

Stepladder: a ladder built with rectangular stiles and treads (not rungs) that are designed to be horizontal in use (*see also* Steps).

Steps (or pair of steps): a stepladder with a framed stay hinged to the top to make it self-supporting.

String (or stringer): a sloping board at each end of the treads that carries the treads and risers of a stair.

Tread: the (usually) horizontal surface of a step; also the length (from front to back) of such a surface.

Turret step: a triangular stone step from which a spiral stair is built up.

Vis (or vice or Vis de Saint Gilles): a helical stair.

Wash (or kilt): a slight sloping of treads to throw off rainwater.

Well: an open space through one or more floors.

Winder (or wheel step): a tread of triangular or wedge shape (*see also* Balanced step).

Winding stair: a spiral stair; a circular or elliptical geometrical stair.

References

Alberti, Leone Battista. 1955. *Ten Books on Architecture, Book 1, 1485.* London: Tiranti.

Alberti, Leone Batista. 1986. *The Ten Books of Architecture: The 1755 Leoni Edition.* Book 1, chapter XIII. New York: Dover Publications.

Anderson, William. 1970. *Castles of Europe.* London: Paul Elek.

Bachelard, Gaston. 1969. *The Poetics of Space.* Boston: Beacon Press.

Bek, Lise. 1985. "The Staircase and the Code of Conduct." In André Chastel and Jean Guillaume, eds., *L'Escalier.* Paris: Picard.

Blondel, François. 1675–1683. *Cours d'Architecture Enseigné dans l'Académie Royale d'Architecture.* Paris: Lambert Roulland.

Camesaca, Ettore, ed. 1971. *History of the House.* New York: Putnam.

Connaissance des Arts 74 (1958). "L'Escalier des Ambassadeurs."

Coulton, J. J. 1977. *Ancient Greek Architects at Work.* Ithaca: Cornell University Press.

Flaubert, Gustave. 1954. *Dictionary of Accepted Ideas.* Translated by Jaques Barzun. Norfolk, Conn.: New Directions Book.

Foucault, Michel. 1973. *The Order of Things: An Archaeology of the Human Sciences.* New York: Vintage Books.

Fraser, Douglas. 1968. *Village Planning in the Primitive World.* New York: Braziller.

Freud, Sigmund. 1948. *The Dream-Work: Representation by Symbols in Dreams.* In *The Complete Psychological Works,* vol. 5. London: Hogarth Press.

Giedion, Sigfried. 1964. *The Eternal Present.* New York: Pantheon Books.

Giersch, Ulrich. 1983. "On Steps." *Daidalos* 9.

Gillies, Linda Boyer. 1972. "An Eighteenth Century Roman View: Panini's Scalinata della Trinità dei Monti." *Museum of Modern Art Bulletin* (February–March).

Guadet, Julien. N.d. *Eléments et Théorie de l'Architecture.* 4th ed. Paris: Libraire de la Construction Moderne.

Hugo, Victor. 1947. *The Hunchback of Notre-Dame.* New York: Dodd, Mead.

Kempers, A. J. Bernet. 1959. *Ancient Indonesian Art.* Cambridge, Mass.: Harvard University Press.

Kodre, Helfried. 1983. "Functional Structures in a Historicist Guise: Nineteenth Century Staircase Halls." *Daidalos* 9.

Laing, Alastair. 1978. "Central and Eastern Europe." In *Baroque and Rococo: Architecture and Decoration.* Edited by Anthony Blunt. New York: Harper & Row.

McAndrew, John. 1980. *Venetian Architecture of the Early Renaissance.* Cambridge, Mass.: MIT Press.

Marder, Tod A. 1980. "Bernini and Benedetti at Trinita dei Monti." *Art Bulletin* 62 (June).

Martini, Francesco di Giorgio. 1967. *Trattati di architettura, ingegneria e arte militare* (c. 1480), facs. ed. Milan: Maltese.

Marwick, Thomas Purves. 1888. *The History and Construction of Staircases*. Edinburgh: J. & J. Gray.

Masson, Georgina. N.d. *Italian Gardens*. New York: Harry N. Abrams.

Meyer, C. E. 1967. "The Staircase of the Episcopal Palace at Wurzburg." Ph.D. dissertation, University of Michigan.

Mielke, Friedrich. 1966. *Die Geschichte der Deutschen Treppen*. Berlin: Wilhelm Ernst.

Miller, Henry. 1989. *The Tropic of Cancer*. London: Grafton.

Moore, Charles, Gerald Allen, and Donlyn Lyndon. 1974. *The Place of Houses*. New York: Holt, Rinehart and Winston.

Morgenthaler, Fritz. 1970. "The Dogon People." In Charles Jenks and George Baird, eds., *Meaning in Architecture*. New York: George Braziller.

Muybridge, Eadweard. 1955. *The Human Figure in Motion*. New York: Dover.

Newman, Oscar. 1972. *Defensible Space*. New York: Macmillan.

Norberg-Schulz, Christian. 1971. *Existence, Space and Architecture*. New York: Praeger.

Nydegger, W. F., and C. Nydegger. 1966. *Tarong: An Ilocos Barrio in the Philippines*. New York: Wiley.

Pagels, Elaine. 1988. *Adam, Eve, and the Serpent*. New York: Random House.

Palladio, Andrea. 1965. *The Four Books of Architecture* (1570). Isaac Ware translation of 1738. New York: Dover.

Pevsner, Nikolaus. 1968. *An Outline of European Architecture*. 7th ed. Harmondsworth, England: Penguin.

Picard, Gilbert. 1968. *Roman Painting*. Greenwich, Conn.: New York Graphic Society.

Ponti, Gio. 1960. *In Praise of Architecture*. New York: C. W. Dodge Corp.

Rapoport, Amos. 1969. *House Form and Culture*. Englewood Cliffs, N.J.: Prentice-Hall.

Robertson, D. S. 1969. *Greek and Roman Architecture*. 2d ed. Cambridge: Cambridge University Press.

Rothery, Guy Cadogan. N.d. *Staircases and Garden Steps*. New York: Stokes and Co.

Rumi, Jalalu'ddin. 1982. *The Mathnawi*. Book V. Edited and translated by Reynold A. Nicholson. Cambridge: Cambridge University Press.

Scamozzi, Vincenzo. 1615. *L'Idea della Architettura Universale*. Vol. 2. Venice.

Scully, Vincent. 1969. *American Architecture and Urbanism*. New York: Praeger.

Sohm, Philip L. 1985. "The State and Domestic Staircase in Venetian Society and Politics of the Renaissance." In André Chastel and Jean Guillaume, eds., *L'Escalier*. Paris: Picard.

Thacker, Christopher. 1979. *The History of Gardens*. Berkeley: University of California Press.

Théâtre National de l'Opéra de Paris. N.d. Paris: Publications de l'Opéra de Paris.

Vasari, Giorgio. 1946. *Lives of the Artists*. 1550 edition, abridged and edited by Betty Burroughs. New York: Simon and Schuster.

Vitruvius (Marcus Vitruvius Pollio). 1926. *The Ten Books on Architecture*. Translated by M. H. Morgan. Cambridge: Harvard University Press.

Vollmer, Emil. N.d. *Instituto Geografico de Agostino*. New York: Reynal and Company.

Whiteley, Mary. 1985. "La Grande Vis." In André Chastel and Jean Guillaume, eds., *L'Escalier.* Paris: Picard.

Wilkinson, Catherine. 1975. "The Escorial and the Invention of the Imperial Staircase." *Art Bulletin* 57 (March).

Wittkower, Rudolf. 1986. *Art and Architecture in Italy, 1600–1750.* Harmondsworth, England: Penguin.

Wotton, Sir Henry. 1686. *The Ground-Rules of Architecture, Collected from the Best Authors and Examples.* London.

Yates, Frances A. 1966. *The Art of Memory.* Chicago: University of Chicago Press.

Yeats, William Butler. 1966. "Blood and the Moon." In *The Winding Stair and Other Poems. Variorum Edition of the Poems of W. B. Yeats.* Edited by Peter Allt and Russell K. Alspach. New York: Macmillan.

INDEX

Aalto, Alvar, 88
   Baker House dormitory, 88
Aghios Petros, 68
Alberti, Leone Battista, x, 72, 88, 171
Albini, Franco, 167
Alessi, Galeazzo, 34, 100
Ammanati, Bartolomeo, 105
Athens, 47
   Parthenon, 44, 49
Avignon, Palace of the Popes, 89
Axis and stairs, 23, 38, 97–100, 103, 108, 113, 116, 124, 131, 133, 135, 136, 140

Baalbek, 49, 120
Bachelard, Gaston, 10
Bacon, Francis, 53, 94
Balustrades, 23, 28, 94, 133, 136, 153, 169, 173–174
Barabudur, 38
Baroque stairs, 7, 11, 25–26, 32, 45, 49, 51, 59, 72, 76, 81, 90, 92, 98–99, 102–103, 105, 108, 111, 113, 116, 121–122, 124, 127–128, 130, 135–136, 140, 142, 145, 150, 160–161, 164, 171
Bergamasco, Giovanni Castello, 100
Berlin, 81
   Altes Museum, 163
   Palace of Prince Albrecht, 140
Bernini, Gian Lorenzo, 25–26, 60, 130, 148, 150, 171
   Scala Regia, Vatican, 25
Bianco, Bartolomeo, 34, 122
Bibiena, Giuseppe, 140, 142
Blois, Château, 34, 60, 64
Blondel, François, x, 42

Bofill, Ricardo, 111
Borromini, Francesco, 150
   Barberini Palace, Rome, 74
Braga, Bom Jesus do Monte, 45
Bramante, Donato, 60
   Cortile del Belvedere, Vatican, 68, 103, 116
Bruchsal, Episcopal Palace, 102, 127, 130, 150
Bruhl, Schloss Augustusburg, 102, 130, 136
Brunelleschi, Filippo
   Foundling Hospital, Florence, 90

Caprarola, Farnese Palace, 68, 105–106, 145
Caserta, Royal Palace, 108, 127
Cathedrals, 45–46, 49, 54, 60
Chambord, Palace, 76, 120
Climbing poles, 10, 13, 14, 18, 19, 32, 34, 38, 46, 51, 76, 128, 150
Cluny, abbey church, 54
Codes, building and fire, xi–xii, 163, 164
Coeur, Jacques, 120
Companionways, 18, 19, 51
Composite stairs, 68, 87
Cordonata, 3, 49, 173
Covarrubias, Alonso de, 94, 97, 98, 100, 171
Crepidoma, 32, 44, 113

Debussy, Claude, 164
Defensive stairs, 14, 54
Delos, 89
Dientzenhofer, Kilian Ignaz, 131
Dimensional inconsistency, 23
Dogleg stairs, 11, 87, 89, 90, 92, 94, 97, 103, 155, 167
Double helix stairs, 74, 76
Double-riser stairs, 19
Du Cerceau, Jacques, 145

Egas, Enrique, 94
Elevators, xi, 161, 164, 167, 169
Escalators, xi, 89, 161, 164, 167, 169
Escorial. *See* Madrid

Falls, x, xii, xiii, 26, 55, 88, 108, 136
Fischer von Erlach, Johann Bernhard, 130, 133
Florence
   Baptistery, 46
   Foundling Hospital, 90
   Laurentian Library, 13, 32, 49, 119, 122, 128, 145, 161
Fontainebleau, Palace, 145–146
Foucault, Michel, x, 16
Francesco di Giorgio Martini, x, 44, 94, 99, 124, 128
Freud, Sigmund, 10

Gait, ix, x, xiii, 5, 23, 42, 44, 130
Garden stairs, 102, 105, 111, 113, 116
Garnier, Jean-Louis-Charles, 145, 164
Genoa, 34, 97, 98, 100, 122, 153
   Palazzo Doria, 100
   Palazzo Marcello-Durazzo, 34, 122
   Palazzo Municipio, 34, 100, 122
   Palazzo Rosso, 167
Giedion, Sigfried, 34
Goethe, Johann Wolfgang von, 28
Gothic stairs, 28, 60
Gradient, 18, 23, 174
Grand approach stairs, 87, 116
Graz, 76, 130, 153
Gropius, Walter, 60
Guadet, Julien, x, 122, 128, 145
Guarini, Guarino, 150
Guas, Juan, 94

Handrails, 18–19, 23, 25, 68, 97, 133, 145, 173, 174
Hardwick Hall, 26
Hatfield House, 94
Hazards, xii, xiii
Hedingham Castle, 56
Heilbrunn, Kilianskirche, 57, 64

Helical stairs, 11, 26, 34, 53–81, 87, 89–90, 92, 94, 124, 142, 145, 150, 153, 155, 158, 167, 169, 171, 173–175
Herrera, Juan de, 98–100
Hildebrandt, Johan Lucas von, 130–135, 136, 140
   Belvedere Garden Palace, Vienna, 130, 133–135
   Daun-Kinsky Palace, Vienna, 130, 131–133
   Mirabell Palace, Salzburg, 130, 133, 140
   Schloss Weisenstein, Pommersfelden, 102, 130–131, 140, 150
Horta, Victor, 92
   Paul-Emile Janson house, 92
Hugo, Victor, 45, 150, 164

Illumination, x, 7, 10, 23, 26, 57, 60, 66, 68, 74, 92, 94, 113, 131, 136, 140, 150
Imperial stairs, 49, 87, 92, 98–100, 102, 103, 124, 128, 133, 136, 140, 145, 150, 153
Indonesia, 34, 38

Jacobean stairs, 94
Jacobsen, Arne, 60, 167
   town hall, Rodovre, 167
Jones, Inigo, 94
Juvarra, Filippo, 140

Knossos, 94

Ladders, xi, 3, 7, 10, 13–16, 18–19, 36, 51, 111, 122, 173–175
   rungs, 14, 18, 175
   siege, 16, 76
Landings, 13, 26, 32, 38, 55, 87–89, 94, 98–99, 113, 124, 128, 130–131, 135, 136, 140, 142, 145, 150, 153, 155, 157–158, 173–175
Le Brun, Charles, 128
Le Corbusier, 3, 57
   Villa Savoye, 30
Leonardo da Vinci, x, 34, 46, 60, 76, 92, 113, 116
Le Vau, Louis, 23, 128
Light. *See* Illumination
Ligorio, Pirro, 103, 105–106, 116, 171

Linz, St. Florian, 153
Lucullan gardens, 103
Lurago, Rocco, 34, 100, 171
Lutyens, Sir Edwin, 19
Lyminge, Robert, 94

Maderno, Carlo, 108
Madrid, 98, 100
   Alcazar, 98–100, 128
   Escorial, 99–100, 102
Martini. *See* Francesco di Giorgio Martini
Marwick, Thomas Purves, 47
Masson, Georgina, 110
Medieval stairs, 7, 11, 16, 28, 32, 44–46, 49, 54–57, 59, 64, 81, 92, 102–103, 113, 120
Mexico, 5, 14
Michelangelo Buonarroti, 34, 116
   Campidoglio, Rome, 49–51, 116–119
   Laurentian library, Florence, 13, 32, 49, 119, 122, 128, 145, 161
Mielke, Friedrich, 44, 74, 163
Miller, Henry, 10
Modernism, 164, 167, 169
Monumental stairs, 23, 47
Moore, Charles, 26
Munich
   Palace of Justice, 161, 167
   State Library, 160
Myth, 7

Naples, 32, 131, 153–157. *See also* Sanfelice, Ferdinando
Nash, John, 66
Neumann, Balthazar, 130, 135–136, 140, 150, 153, 160
   Episcopal Palace, Bruchsal, 102, 127, 136, 150
   Episcopal Palace, Wurzburg, 102, 127, 130, 136, 150
   Hofburg palace, Vienna, 140, 160
   Schloss Augustusburg, Bruhl, 102, 130, 136
Neutra, Richard, 68
Newels, 54–56, 64, 66, 68, 81, 87, 92, 94, 97, 135, 171, 174–175
New York, 18, 42, 44
Niemeyer, Oscar, 66

Norberg-Schulz, Christian, 34
Nordlingen, Georgenkirche, 66
Norman stairs, 16, 55, 56

Orford Castle, 55
Orvieto, 76

Padula, Carthusian monastery, 157
Palladio, Andrea, x, xi, 49, 60, 72, 76, 88, 142
Paris, 57, 76, 167
   Conciergerie, 57, 64
   La Villette park, 124
   Louvre, 57, 150
   Notre Dame, 19, 45, 54, 64, 164
   Opera House, 145, 164
   Pompidou Center, 28, 29
Penshurst Place, 57
Persepolis, 47
Pevsner, Nikolaus, 25, 99, 122, 150
Piranesi, Giambattista, 10
Pisa
   Cathedral, campanile, 28, 60, 64, 68, 113
   San Niccola, 68
Podia and platforms, 45, 46, 113
Pommersfelden, Schloss Weisenstein, 102, 130–131, 140, 150
Pompeii, 26, 116
   amphitheater, 7, 108, 116, 121
Ponzello, Domenico and Giovanni, 100
Porta, Giacomo della, 108
Portman, John, 64
Praeneste (Palestrina), 103

Ramps, xi, 3, 13, 19, 23, 28, 30, 36, 60, 68, 76, 102–103, 105–106, 122, 133, 145, 167, 175
   stepped ramps, 3, 5, 19, 32, 38, 44–46, 49, 105–106, 111
Rapoport, Amos, 34
Ravenna, San Vitale, 53
Renaissance stairs, 11, 28, 32, 34, 45, 49, 57, 68, 74, 76, 81, 89, 90, 92, 94, 100, 102–103, 105, 111, 113, 116, 121–122, 131, 145, 171
Risers, x, 5, 13, 19, 23, 25, 28, 32, 34, 42, 44, 66, 88, 121, 127, 130, 173–175
   dimensions, 5, 34

Rogers, Richard, 28
Rome, 34, 102, 130, 140, 148. *See also* Vatican
　Barberini Palace, 74
　Campidoglio, 49–51, 116–119
　Colosseum, 25
　Farnese Palace, 90
　Lateran Palace, 45, 53
　Ripetta steps, 34, 148
　St. Peter's, 44
　Santa Maria in Aracoeli, 87
　Santa Maria Maggiore, 46
　Spanish Steps, 34, 103, 148
　Trajan's Column, 53
　Villa di Papa Giulio, 105, 145
Rothenburg, 60
Rothery, Guy Cadogan, 3, 44, 55, 94
Rumi, Jalalu'ddin, 7
Rungs, 14, 18

Salamanca, Colegio de Los Irlandeses, 94, 97
Salzburg, Mirabell Palace, 130, 133, 140
Sanfelice, Ferdinando, 76, 153–158
　Palazzo Bartolomeo di Maio, 155, 157
　Palazzo Capuano, 157
　Palazzo Sanfelice, 153
　Palazzo Serra di Cassano, 155
Sangallo, Antonio da, 76
　Farnese Palace, Rome, 90
Scale and stairs, 32, 36, 42, 45, 49, 51, 99, 108, 128, 161, 167
Scamozzi, Vincenzo, x, 90
Schinkel, Karl Friedrich, 81, 140
　Altes Museum, Berlin, 163
　Palace of Prince Albrecht, Berlin, 140
Semper, Gottfried
　Vienna City Theater, 163, 164
Siloe, Diego de, 94
Slips and slip resistance, xiii, 3
Spalato, Palace of Diocletian, 120
Spiral stairs. *See* Helical stairs
Staircases of honor, 113, 122, 128
Stairs of the Gods, 34–41
Stairwells, ix, xi, xiii, 5, 26, 42, 49, 55, 59, 60, 68, 72, 74, 76, 87, 90, 92, 94, 97–100, 106, 122, 133, 148, 150, 157, 164, 167, 173–175

Steiner, Rudolph, 92
Sterling, James
　Stuttgart art gallery, 81, 124
Straight flight stairs, 5, 11, 13, 19, 26, 32, 34, 51, 59, 64, 68, 74, 76, 81, 87–89, 94, 99, 113, 122, 131, 142, 145, 148, 155, 158, 160, 174
Strasbourg
　cathedral, 60
　Frauenhaus, 68
Stupas, 38
Stuttgart, Schloss Solitude, 145
Symbolism of stairs, x, 16, 34, 38, 44, 47, 59–60, 88, 103, 108, 121, 160, 167
Symmetrical facades, 60, 116

Tivoli. *See* Villa d'Este
Toledo, 94–100
　Alcazar, 98–100, 128
　Hospital of San Juan Bautista, 90, 94, 97–98
　Mendoza Hospital, 94, 97
　Palace of La Calahorra, 94
　San Juan de Los Reyes, 94, 97
Treads, x, 3, 5, 13, 18–19, 23, 25, 28, 32, 34, 42, 44, 55–56, 64, 68, 74, 121, 127, 130, 145, 150, 173–175
Treppenhaus, 32, 140
Tschumi, Bernard, 124
Tudor stairs, 94
Turbino, Antonio, 142
Turin, Palazzo Carignano, 150

Uxmal, 38

Vanvitelli, Luigi, 127
Vasari, Giorgio, x, 46, 68, 90, 105
Vatican
　Belvedere, 60, 68, 103, 116, 122, 130, 133, 135–136
　Scala Regia, 25–25
Vaux-le-Vicomte, 23
Venice, 72, 74, 90, 113
　San Giorgio Maggiore, 128
　Scala dei Giganti, 113
　Scala d'Oro, 90, 113, 171

Versailles, 128
Vienna, 130–131
   Belvedere Garden Palace, 133–135
   City Theater, 163
   Daun-Kinsky Palace, 131–133
   Hofburg palace, 140, 160
   Palace of Justice, 161
Vignola, Giacomo Barozzi da, 60
   Farnese Palace, Caprarola, 68, 105–106, 145
   Villa Lante, Bagnaia, 106
Villa Aldobrandini, Frascati, 108
Villa Bettoni, Bogliaco, 111
Villa Capra, Vicenza, 49
Villa Corsini, Mezzomonte, 108
Villa d'Este, Tivoli, 103, 105–106, 122, 145, 171
Villa Foscari, Malcontenta, 142
Villa Garzoni, Collodi, 111
Villa Lante, Bagnaia, 106
Villa Lechi, 142
Villa Medici, Fiesole, 103
Villa Torrigiani, Siena, 110
Viollet-le-Duc, Eugène-Emmanuel, 164, 171
Vis. *See* Helical stairs
Viso del Marques, 100
Vitruvius Pollio, Marcus, x, 23, 44

Water stairs, 103, 105–106, 108, 110, 124
Well. *See* Stairwells
Wittkower, Rudolf, 127–128, 171
Wood, John, 158
Wooden stairs, 16, 76, 94
Wright, Frank Lloyd, 28
   Guggenheim Museum, New York, 72
Wurzach Palace, 158
Wurzburg
   Episcopal Palace, 102, 127, 130, 136, 150

Yates, Frances A., 7